ANALYSIS OF
AIR POLLUTANTS

ANALYSIS OF AIR POLLUTANTS

PETER O. WARNER
WAYNE COUNTY DEPARTMENT OF HEALTH
AIR POLLUTION CONTROL DIVISION

TD
890
.W36

A WILEY-INTERSCIENCE PUBLICATION

JOHN WILEY & SONS

NEW YORK • LONDON • SYDNEY • TORONTO

Copyright © 1976 by John Wiley & Sons, Inc.

All rights reserved. Published simultaneously in Canada.

No part of this book may be reproduced by any means, nor transmitted, nor translated into a machine language without the written permission of the publisher.

Library of Congress Cataloging in Publication Data:

Warner, Peter O   1937–
 Analysis of air pollutants.

 (Environmental science and technology)
 Includes bibliographical references.
  1. Air—pollution—Measurement.   2. Air sampling apparatus.   I. Title.

TD890.W36 628.5'3 75-26685
ISBN 0-471-92107-6

Printed in the United States of America

10 9 8 7 6 5 4 3 2 1

SERIES PREFACE
Environmental Science and Technology

The Environmental Science and Technology Series of Monographs, Textbooks, and Advances is devoted to the study of the quality of the environment and to the technology of its conservation. Environmental science therefore relates to the chemical, physical, and biological changes in the environment through contamination or modification, to the physical nature and biological behavior of air, water, soil, food, and waste as they are affected by man's agricultural, industrial, and social activities, and to the application of science and technology to the control and improvement of environmental quality.

The deterioration of environmental quality, which began when man first collected into villages and utilized fire, has existed as a serious problem under the ever-increasing impacts of exponentially increasing population and of industrializing society, environmental contamination of air, water, soil, and food has become a threat to the continued existence of many plant and animal communities of the ecosystem and may ultimately threaten the very survival of the human race.

It seems clear that if we are to preserve for future generations some semblance of the biological order of the world of the past and hope to improve on the deteriorating standards of urban public health environmental science and technology must quickly come to play a dominant role in designing our social and industrial structure for tomorrow. Scientifically rigorous criteria of environmental quality must be developed. Based in part on these criteria, realistic standards must be established and our technological progress must be tailored to meet them. It is obvious that civilization will continue to require increasing amounts of fuel, transportation, industrial chemicals, fertilizers, pesticides, and countless other products and that it will continue to produce waste products of all descriptions. What is urgently needed is a total systems approach to modern civilization through which the pooled talents of scientists and engineers, in cooperation with social scientists and the medical profession, can be focused on the development of order and equilibrium to the presently disparate segments of the human environment. Most of the skills and tools that are needed are already in existence. Surely a technology that has created such manifold environment problems is also capable of solving them. It is our hope that this Series in Environmental Sciences and Technology will not only serve to make this challenge more explicit to the established professional but that it also will help to stimulate the student toward the career opportunities in this vital area.

Robert L. Metcalf
James N. Pitts, Jr.
Werner Stumm

PREFACE

In the recent past, much of the data that have characterized the environmental deterioration of our atmosphere have suffered from limitations inherent in applying, without adaptation, classical chemical methodology to the chemical analysis of air pollutants. Only recently, through the efforts of the U.S Environmental Protection Agency, together with other municipal agencies both here and abroad, have the necessary skills been consolidated to arrive at analytical methods to determine trace but nonetheless hazardous quantities of pollutants in air.

This text has been assembled in response to the need to present this information in an organized form, and is addressed to undergraduate science majors and beginning graduate students as both a program of study and a reference manual in collecting and analyzing air samples, primarily in the urban environment. Some discussion is devoted to the origin and properties of common and some less-common air contaminants. At the beginning, considerable emphasis is placed on methods for the identification of dust-particulates, since these form the largest, most visually obvious class of air contaminants.

Organic compounds, which find their way into the urban industrial atmosphere, are treated from the point of view that physical separation must be adapted to the sampling method if a specific compound is to be determined selectively. Alternatively, advantages are presented for the determination of organic compound families such as total aldehydes, or the index of total aromatics, since these chemically relatable groups are often traceable to a common community source.

Later, in the treatment of gaseous air contaminants, the air chemistry of the individual gases is discussed in some detail, especially when this knowledge is helpful in understanding the principle as well as the application of a given method of analysis. Occasionally, for example, in the case of monitoring ambient hydrocarbons, the object of a given air quality measurement may be the control of some other pollutant, in this case, total oxidant, whose presence is ultimately related to the air chemistry of the nonmethane portion of hydrocarbons.

Since the result of the chemical analysis is only as reliable as the collection method used to obtain the air sample, a great deal of attention has been devoted to methods of air sampling. Here I have encouraged the reader to select the best method of air sampling by weighing advantages and disadvantages of a number of sampling approaches in various situations. I emphasize not only the mechanics of separating the contaminant sample from the body of air, but also the need for accuracy in the measurement of the body of air itself, so that actual concentration of pollutant in amount per unit air volume can be analytically significant.

A growing awareness of the need for reliable measurements of community odor has prompted an extensive review of up-to-date methodology in sensory-response analysis. The theoretical basis for these measurements is presented together with detailed procedures and equipment needed to arrive at meaningful reproducible odor intensities.

Many such step-by-step procedures are included together with simplified directions designed to enable the beginning environmentalist with a background in laboratory science to set up sampling equipment and collect useful data. On the other hand, more sophisticated approaches such as spectrofluorescence and x-ray diffraction are described in some detail to encourage the use of these instrumental methods, which furnish almost invaluable aid to the air pollution chemist in specifically identifying urban-industrial pollutants, such as polynuclear aromatic hydrocarbons, and minerals, such as silica and cermets, which recent medical research indicates are related to human health.

PETER O. WARNER

Detroit, Michigan
July 1975

ACKNOWLEDGMENTS

I wish to express deepest gratitude for the help extended by Dr. James O. Jackson and Mr. Lawrence Saad for their work in outlining and testing a number of the analyses and calibration procedures described in this text, and to Mr. Alan Greenberg for his help in reviewing material related to meteorology and statistics.

Appreciation is also expressed for the very helpful suggestions offered by Dr. Ralph Smith of the University of Michigan, Mr. Burr French of the Ethyl Corporation and Mr. Robert Bower of the Wayne County Air Pollution Control Agency in reviewing individual chapters.

P.O.W.

CONTENTS

1. Origin and Identification of Particulate Air Pollutants — 1
2. Sources and Measurement of Organic Air Contaminants — 55
3. Sources and Measurement of Inorganic Air Contaminants — 95
4. Continuous, Automated Methods of Air Analysis — 166
5. Principles of Air Sampling — 196
6. Calibration of Sampling Instruments and Preparation of Standard Gas Mixtures — 271
7. Odor Detection and Determination — 289

Index — 321

ANALYSIS OF
AIR POLLUTANTS

1

ORIGIN AND IDENTIFICATION OF PARTICULATE AIR POLLUTANTS

INTRODUCTION

The solution to many problems encountered in the identification of airborne particulate matter lies in the synthesis of a considerable amount of knowledge gathered from chemistry, physics, microscopy, crystallography, and ultimately experience.

It is estimated that up to 15% of the total settleable dust and perhaps as much as 25% of suspended particulate is of natural origin. Thus pollution itself is by no means pure pollution. Its composition is governed not only by the makeup of the industry that causes it, but also by the makeup of the land on which the industry stands. The Western states are noted for siliceous air, which is made up of fine particles whose settleable air loading varies from 3 to 5 tons of sand per square mile per month. Air masses in the arctic are noted for "ice fogs," which form at temperatures below 45°F as a result of humidity rising from both natural sources and population centers scattered throughout the icy wasteland. And marine and coast atmospheres contain suspended salt nuclei. Because of these symbiotic conditions in natural air, it is well to consider the earth's natural atmosphere before exploring its sometimes less natural contaminants. Among the natural contaminants that make up a portion of any collected pollutant samples are suspended dust from volcanoes, soil from air erosion of farmland, hydrocarbons from coniferous vegetation, pollen, and particulate carbon smoke of natural origin.

If we assume an air loading of 100 $\mu g/m^3$—as exists in many urban atmospheres—it is surprising to find 45% silica, 13% Fe_2O_3, 6% Al_2O_3, 24% carbon, 1% lime, 1% alkali sulfates, and 3% limestone, and the balance a mixture of such materials as vanadium, magnesium, alkali carbonates and chlorides, titanium, zincite, and a number of light and heavy metal sulfates together with organics and fibers.

COMMON AIR CONTAMINANT PARTICULATES

Probably more time, effort, and money have been spent on the study of fuel combustion products as air contaminants than on any other source in industrial technology. The introduction of the automobile added only one new class of particulates to the queue of classes awaiting their turn as potential hazards to human health and as a source of consternation to the laboratory chemist working at their separation and identification. Particles such as lime, limestone, and cement dust from kilning operations, coke dust and polycyclic aromatics from coking operations, iron oxides from the various ore smelting procedures, and fluorides from nonferrous metallurgy have joined with asphalts, solvents, synthetic monomers, butyl rubber, and carbon black from the construction and heavy polymer industries. All these emissions are homogenized with such proprietary air species as fly ash from electrical power plants and windblown slag from storage piles in the vicinity of blast furnaces and the road construction material process plants, to which the iron and steel industries address this marketable by-product. Blown insulation is still another example of a proprietary air contaminant.

Home space heating is perhaps the least major source of carbonaceous particulate, although such urban materials as windblown salt from winter treatment of roads have recently appeared as rather substantial causes of extensive local particulate pollution. Further examples of such local pollution are the following:

1. Particles of copper, copper oxides, boron and silicon carbides, iron and iron oxide, aluminum, magnesium, and silica from grinding and milling operations.

2. Carbon black and kish (graphitic carbon) graphite from steel and rubber processing.

Of the major particulate pollutants probably the most prevalent and very often the most difficult to separate from road dust and natural pollutants is common fly ash. The discussion that follows should help to delineate the classification of types of fly ash contaminant and aid in its separation and identification in a typical sample of industrial particulate.

Coal and Coal Fly Ash

The most common sampling techniques for particulates (as described in Chapter 5) involve dustfall and high-volume sampling. These produce sample weight in excess of 1 g and particles in the overall range of 0.1 to 100 μ in diameter. This is a good sample analytically, since most analytical methods involving the determination of metals, for example, begin with samples of particle size similar to that of settleable dust. In this case, the samples are already rather finely divided and require little sample preparation except for the high-volume filtered dust that must itself be removed from the filter medium before analysis. This is usually accomplished by dissolution in dilute nitric or hydrochloric acid or by ultrasonic peptization if such instrumentation is available.

The chemistry used in separating the collected particles depends on the nature of the sample. Combustion is a very complex process, so that fly ash may proceed through a number of mechanisms, giving particles of various types. Typical fly ash (Figure 1) looks like a mixture of geometrically amorphous, but surprisingly characteristic, particles whose approximate shape, particle diameter, and relative degree of agglomeration permit quite reliable optical identification.

When coal burns, the heat generated may vaporize material that can subsequently condense to produce the fine particles that are called soot. These range in size from less than 0.1 to 10.0 μ in diameter. Depending on the volume of air present and the temperatures involved, the burning of carbon fuel may yield particles as small as 50 Å—a class with properties similar to those of the pyrophoric powders of refractory metals, or the more familiar face talc.

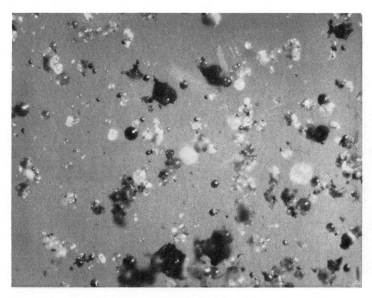

Figure 1. Typical coal fly ash. (Magnification 100×.)

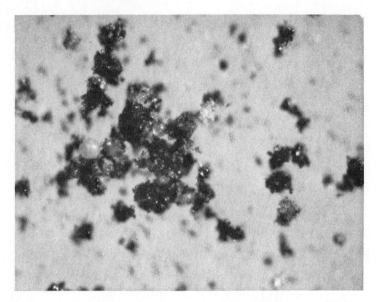

Figure 2. Coal, fly ash mixed with fine carbon black. (Magnification 100×.)

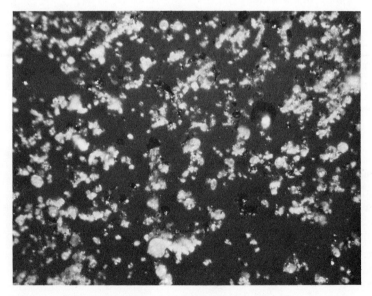

Figure 3. Pulverized coal fly ash. (Magnification 100×.)

Common Air Contaminant Particulates 5

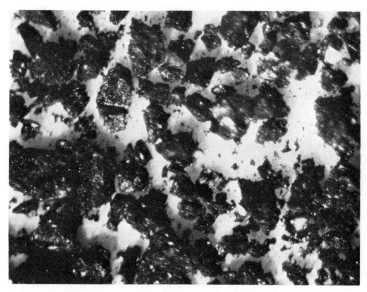

Figure 4. Windblown coal particulate. (Magnification 40×.)

Such particles of carbon are, as one might imagine, highly surface-active and rather graphitic in appearance under microscopic examination.

Ordinarily, very small particles ($<0.1\ \mu$) are quite unstable under the gross combustion conditions usually associated with space heating and industrial power plant operations. Hence they are less common in the average urban particulate. One important exception is the submicron carbon black that escapes from automobile tires through ordinary road wear. This "dry grinding" often results in the unzipping of the butadiene-styrene polymer system with substantial formation of monomer, together with carbon black, which is thought to be quite long-lived in the atmosphere. Figure 2 illustrates this particulate and shows its relative size when mixed in equal portions with a sample of ordinary fly ash.

If the fuel itself is particulate, such as fuel designed for pulverized coal-fired boilers, the resulting particulate looks as shown in Figure 3. Here, it is noteworthy that the carbonaceous dust is reduced to practically the size of table salt. This salt is merely the concentrated, mixed alkalies and heavy metal compounds present in the original fuel that have been converted to carbonates, sulfates, and halides in the combustion process.

Nearly all coal combustion produces fly ash of the appearance shown in Figures 1 and 3. Since coal is usually stored near its combustion site, the fly ash found in the immediate to proximate vicinity of coal burning facilities is very

Figure 5. Stoker coal fly ash particulate. (Magnification 40×.)

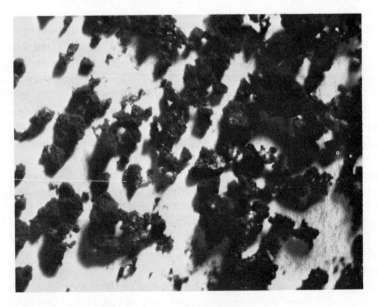

Figure 6. Oil fly ash. (Magnification 40×.)

often mixed with coal particles (Figure 4). These are rather easily recognized by their generally smoother surface, which may in some cases appear dull but oily. Such windblown coal dust seriously increases local air pollution but is often overlooked because of the more obvious emission of carbon as a smoke plume. The particulate produced by stoker coal (Figure 5) is cusped, probably because of the preferential downward direction of burning in the stoker-fired furnace or boiler.

Oil burning produces a characteristic field of particles somewhat smaller than those produced by coal burning. These particles (Figure 6) are remarkably uniform in size distribution—a property that remains seemingly independent of the type of burner used, except when a malfunctioning burner produces a large volume of carbon soot from which very small particles aggregate to form chain or brush-heap-like, low density clumps. Careful probing with a filament may reveal such properties and aid in the identification of coal or oil soot.

Coke and Coke Fly Ash

The burning of coke produces particulate as shown in Figure 7, whereas coke itself (Figure 8) looks glossy and appears, under microscopic examination, to be made up of the following:

1. Coke balls: shiny spheres, especially prominent in petroleum coke.
2. Coke char: flat flakes irregular in shape.
3. Coke glance: shiny flat surfaces in which some cell structure is visible.

A very common mixture is that of particulate coke and coal that arises from air streams over a common storage area such as supply yards or from coking operations that use coal as starting material. Here the degassed coke appears much shinier than does coal, whose surface is coated with amorphous tars. It should be noted that a method exists for separating coke and coal miscroscopically, based on the specific reflectance of a large population of particles.[1] Such solid fuel samples also lend themselves to separation by x-ray diffraction, described later in this chapter.

Other Combustion-Related Particulates

Since many power plants have already been converted to the combustion of natural gas, with marked subsequent improvement in the quality of community air, it is interesting that even this advance in technology may be an occasional source of a product called gas soot, which is unlike most previously discussed

Figure 7. Coke fly ash particulate. (Magnification 40×.)

Figure 8. Coke particulate. (Magnification 40×.)

black particulate in that only a very small quantity of mineral residue is included with the carbon particulate.

Incineration of refuse produces very heterogeneous particulate. The material incinerated largely determines the characteristics of the emission, although there are similarities in particulate collected from a municipal sewage incinerator (Figure 9) and from a refuse incinerator (Figure 10).

Dustfall samples frequently contain particles of windblown soil (Figure 11), which is microscopically similar to fly ash and may be mistakenly identified as such on cursory examination. It is therefore advisable to use signs of fusion as evidence for the presence of fly ash. Figure 1 shows partial encapsulation of carbon granules by the more vitreous siliceous components. Furthermore, because both fly ash and road dust contain large amounts of free silica, evidence of sintering of the silica nodes in the fly ash often sets this component apart from the generally larger and rougher-edged silica found in road dust. However, abrasion and pulverization by surface vehicles may considerably reduce these optical dissimilarities.

Cement Dust

Probably the most prevalent class of mineral pollutants consists of the carbonates and silicates used as construction materials. Although the most common

Figure 9. Municipal sewage incinerator particulate. (Magnification 40×.)

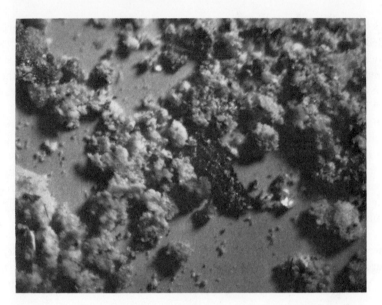

Figure 10. Municipal refuse incinerator particulate. (Magnification 40×.)

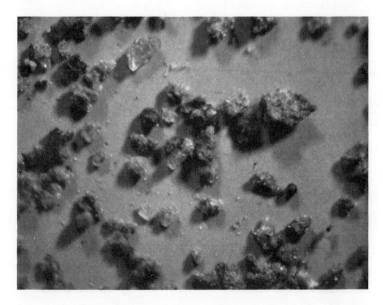

Figure 11. Typical windblown soil particulate. (Magnification 40×.)

Figure 12. Cement product particulate. (Magnification 100×.)

product of a cement plant is kilned calcium pentasilicate ($CaSiO_5$), much of the plant's actual stack emission is calcium carbonate. This results from the actual kilning operation in which components such as lime (very often fines from the kilning of CaO), silica, alumina, and iron oxides are heated together with limestone and various proprietary clays at 2700°F and above to produce the product known as *portland cement*. Blast furnace slag, bentonite, gypsum ($CaSO_4$), and occasionally straw or other fibrous material are added to the premix. Calcium oxide, which aggregates with difficulty, is coemitted with carbon dioxide to form calcium carbonate in the aerosol phase.

Any one of these component particulates is likely to be found in the immediate vicinity of the processing plant because of wind effects on stockpiles, while 1 to 2 miles away only the actual stack or kiln emission prevails.

It is not possible to illustrate each cement-related emission here; the actual product—a mixture of calcium silicates—is shown in Figure 12. Here the grapelike structure of the particles is its chief characteristic. Other methods of identifying cement are discussed in the section on instrumental methods.

Foundry Dust

Emissions are related to metal refining and processing and are most often optically identifiable through the chemistry of the metal, as described in the

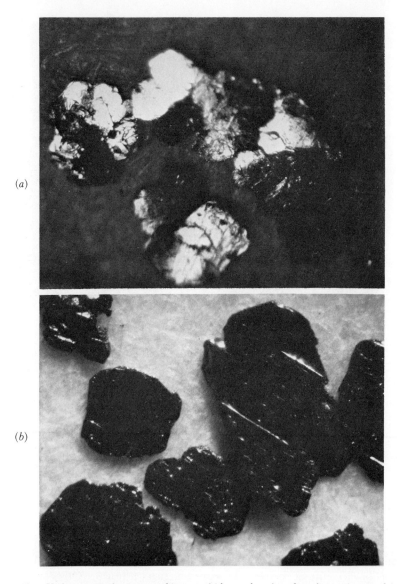

Figure 13. Kish, a typical mixture of iron carbides and carbon found near iron melting operations. (Magnification 40×.) (a) Dark background with overlighting shows a shiny surface. (b) Light background shows grain structure.

next section. Some common metals such as iron oxides from blast and open hearth furnaces are rather easy to recognize, and a typical field of this industrial dust contains partially siliceous reddish yellow particles of the iron oxide. A small hand magnet placed under the field of particles is useful in determining what percentage of the particulate is actually iron or iron oxide related.

Dust from molding sand used in the preparation of metal castings is often found, as is carbon graphitic material called "kish" (Figure 13), which precipitates from molten iron during pouring operations. Other similar emissions include zirconium sand, various clay fillers, and phenolic resin binders used for industrial core molding.

Nonferrous Particulate

Because zinc oxide tends to sublime from the melting of nonferrous alloys during reclaiming operations, a large amount of this very finely divided ($<0.1\text{-}\mu$ diameter) material finds its way into nearly every urban atmosphere. Its small particle size poses problems of collection for both the industrialist and the analytical chemist. Such small airborne particles, which often escape through the chemist's ($<0.1\text{-}\mu$) pore filters, can, however, be collected by means of liquid impingers (Chapter 5). Once collected, the zinc oxide looks as shown in Figure 14.

Mixed Oxide Particulate

The mechanical forming of metal produces still another class of pollutant particle: abrasives and machine grindings.

Abrasive particles are valuable on the basis of their surface hardness and sharp cutting edges, and the same characteristics can be used to recognize both abrasive and polishing substances which are often a source of local pollution. These particles usually are large (10 to 100 μ in diameter) and often contain surface traces of the bonding agent that originally fixed the abrasive grains to a grinding surface. Figure 15 shows particulate silicon carbide, a common milling agent, along with corundum (Al_2O_3) and quartz (SiO_2). Quartz, which is often found near sand blasting operations (Figure 16), is sometimes difficult to distinguish from ground glass (Figure 17). However, the two may be separated easily by x-ray diffraction (quartz produces a Laue crystal pattern), by polarized light using a microscope adapted for cross-polar prisms (quartz is birefringent), or by dispersion liquids; for example, Pyrex glass disappears in a solvent of the same refractive index, such as toluene.

Figure 14. Zinc oxide particulate. (Magnification 100×.)

Figure 15. Silicon carbide grinding abrasive. (Magnification 40×.)

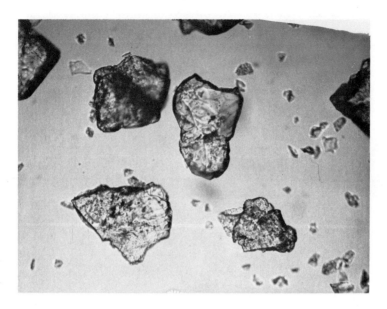

Figure 16. Alpha quartz dust. (Magnification 100×.)

Other Mineral Particulates

Other industrial particulate contaminants include the general class of synthetic detergents, which invariably contain large amounts of washing soda $Na_2SO_4 \cdot 12H_2O$, and the various sequestering agents—trisodium phosphate and sodium isopolyphosphate as well as alkali meta- and orthosilicates. A typical commercial detergent containing sodium aryl sulfonate together with less than 50% sodium isopolyphosphate and sodium sulfate builder appears as shown in Figure 18. In some industrial districts, catalyst dust is a local emission problem. These surface-reactive agents are less difficult to recognize in the presence of fly ash, since "spent" catalyst may appear to be mixed with carbon particles. Petroleum process catalyst looks typically as shown in Figure 19. Such cracking catalysts usually darken in color during their process life and, because of abrasion in fluidized beds, spent particles assume a nearly spherical shape with rather high surface reflectivity.

Airborne Fibers

Of airborne particulates, one rather ubiquitous class of contaminants that is difficult to identify, unless a microscope with polarizing field is available, is

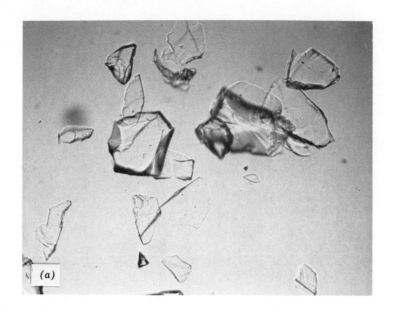

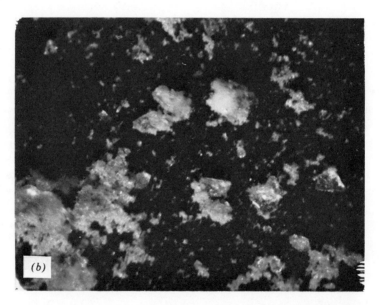

Figure 17. (a) Powdered glass as it appears with underlighting (magnification 100×); (b) Powdered glass as it appears with overlighting. (Magnification of 100×.)

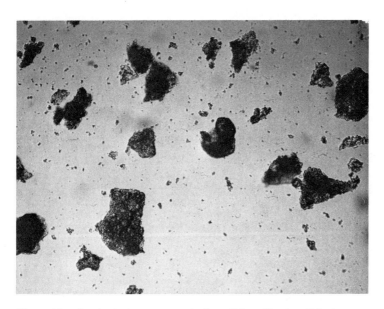

Figure 18. Synthetic detergent particulate. (Magnification 100×.)

Figure 19. Petroleum catalyst particulate. (Magnification 40×.)

18   Origin and Identification of Particulate Air Pollutants

that of fibers. In this category are found various kinds of insulation and packing material, industrial filters, and common "lint," as well as wool and other animal fibers.

Glass fiber (Figure 20) is common in urban and industrial atmospheres both because of its use as an insulator in building construction and because of the widespread use of fiberglass as an air filtering medium.

Mineral fibers such as fiberglass and asbestos can be rather easily separated from the synthetic organic fibers of the rayon and nylon group by heating the microscope slide to cause partial sintering of the lower melting organics. Fiber length is of little help in distinguishing various kinds of natural or synthetic fibers, although a nodelike or scaly appearance often distinguishes animal from nonanimal fibers.

In general, synthetic fibers exhibit a smoother and more uniform surface than do plant fibers without the "flat" side appearance of plant fibers and without traces of cell structure. Wood fibers may also carry traces of sap or resin. Such cellulose fibers will blacken quickly when treated with concentrated sulfuric acid under the field of the microscope. A uniform spreading of the darkening along the leading edge of the acid is due to cell hydrolysis and is a clear indication of cellulose or, a celluloselike synthetic organic product such as rayon or nylon. An indication of styrene or aromatic ring linked polymer fibers

Figure 20.   Glass fibers, windblown insulation. (Magnification 100×.)

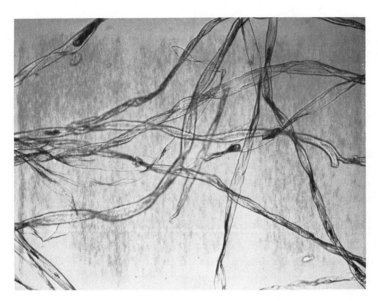

Figure 21. Cotton fibers. (Magnification 100×.)

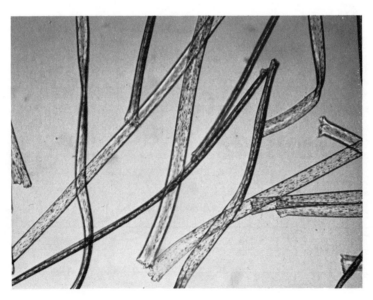

Figure 22. Nylon fibers. (Magnification 100×.)

20   Origin and Identification of Particulate Air Pollutants

in the presence of concentrated H_2SO_4 is the appearance of a reddish or orange color, sometimes changing to brown as the acid converts ring monomers to the corresponding colored aromatic sulfonates. Figures 21 through 24 show photomicrographs of a number of common fiber pollutants that often appear in filtered or settled dust samples. A word of caution to the analyst: the passage of air through some filter material designed to collect industrial pollutants before they reach the outside air may well result in widespread emission of large amounts of fibers if the manufacturer's recommended air flow is exceeded with subsequent rupturing of the collecting surface.

Microscopic Identification by Means of Chemical Preparation

Since the concentration of airborne particulate usually involves filtration, impinging (Chapter 5), or merely settling by gravity (the collection of fallen dust), a surplus of sample is seldom available. Thus it is generally advisable to carry out qualitative chemical reactions under the microscope so as to conserve as much of the sample as possible.

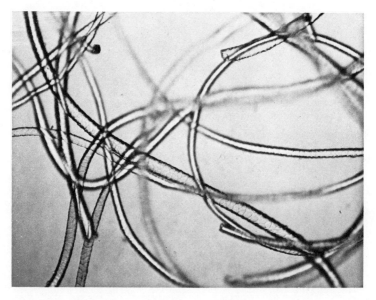

Figure 23.   Wool fibers. (Magnification 100×.)

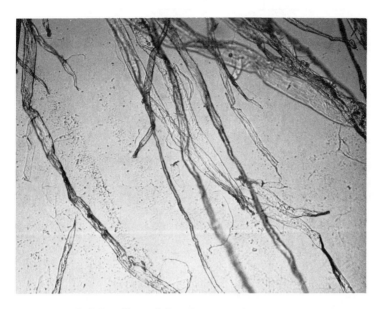

Figure 24. Cellulose fibers. (Magnification 100×.)

Common Chemically Reactive Species

As any chemist knows, there are many spot tests and class reactions that can be used to roughly confirm the presence of a chemical constituent or group of such constituents as ions or radicals by their color formation, effervescence, change in polarity, and so forth. Many of these are available in the works of "Feigl,"[2-4] although still others are more current.[5,6] Many organic compounds are also easy to classify by noting color formation or change in solubility upon the addition of an appropriate reagent. A rack of such reagents is a useful addition to any microscope table; a suggested list of these solutions is given in Table 1, together with the appropriate reaction product and summarized procedure. Although the chemical bases for many such spot tests can be found in the literature, those given in the table are presented here because the species are those most often encountered in air pollution control samples. Where *iron, nickel,* and *copper* are metallic products of smelting, grinding, and alloying, *lead* and *vanadium* are often found as halides resulting from fuel combustion, especially the combustion of gasoline which contains residuals of the vanadium(III) chloride and vanadium(III) oxide reforming catalysts, together with antiknock organic lead, which emerges as the lead bromide and the lead chlorobromide.

Titanium is used as an industrial pigment as well as an abrasive in the form

Table 1 Spot Tests That Can Be Carried Out under the Microscope

Contaminant	Reagent	Reaction
Iron	6 M HCl	Red color, Fe(SCN)$_6{}^{3-}$ stains crystals
Nickel	6 M HCl, neutral with amonia, saturate with 10% sodium acetate solution, add 1% dimethylglyoxime in alcohol	Red precipitate, Ni(DMG) forms, Pdg interferes (Cu forms brownish precipitate)
Titanium	5 M NaOH, neutralize, add H$_2$O$_2$	Yellow color forms from titanyl ion TiO
Copper	3 M HCl, add NH$_4$OH to basic	Blue lake of Cu(NH$_3$)$_4{}^{2+}$
Vanadium	Add 5 M NaOH to vanadate, saturate with H$_2$S	Cherry red solution
Zinc	3 M HCl, add 1% K$_4$Fe(CN)$_6$	Grayish white precipitate Zn$_3$K$_2$[Fe(CN)$_6$]$_2$
Bromide	⎫	Pale yellow precipitate, AgBr
Chloride		White precipitate, AgCl
Nitrite		Pale yellow precipitate, AgNO$_2$
Phosphate		Yellow precipitate, Ag$_3$PO$_4$
Sulfate		White precipitate, AgSO$_4$
Sulfite	⎬ Neutral solution add AgNO$_3$	White precipitate, AgSO$_3$ $\xrightarrow{\text{heat}}$ Ag (black)
Acetate		White precipitate, AgC$_2$H$_3$O$_2$ odor vinegar
Sulfide		Black precipitate, Ag$_2$S
Chromate	⎭	Reddish brown precipitate, Ag$_2$CrO$_4$
Carbonate Bicarbonate	6 M HCl	Effervescence
Mercury as Hg(II)	Copper surface	Black precipitate, Cu, CuO
Aluminum	⎫	White precipitate, Al(OH)$_3$
Chrome	⎬ NH$_4$OH + NH$_4$Cl aqueous	Grayish green precipitate, Cr(OH)$_3$
Iron	⎭	Reddish brown precipitate Fe(OH)$_3$
Magnesium	(NH$_4$)$_2$CO$_3$ in aqueous solution contains 5% ethanol	White precipitate

of its dioxide; *zinc,* released from the molding of "white metal" parts, forms a highly volatile, very finely divided oxide that pervades an air envelope radiating many miles from its source.

Mercury has been found in urban dust in concentrations between 0.11 and 4.0 ppm. Perhaps the greatest single source of this air contaminant is coal combustion, since coal contains up to 5.0 ppm of mercury. In this regard, determination of mercury in power plant stack emission by the mercury-vapor meter method shows this particulate to contain as much ss 2.1 ppm mercury, while bottom ash from the same plant was found to contain 0.04 to 0.06 ppm residual mercury.

Common aerosol alkali metals such as *sodium* and *potassium* that distill from kilning, coking, and sintering operations can be detected in very small quantities (ppm) by the flame test, using platinum wire and hydrochloric acid solution of the air collected sample. *Calcium* and, less frequently, airborne *strontium,* which are briquetted as the fluoride for use as slag forming and fluxing agents, evolve into the atmosphere as halides, which gradually convert to oxide. These are also subject to flame tests; the flame colors are as follows:

Sodium:	yellow.
Potassium:	pink (violet) use cobalt blue glass.
Calcium:	brick red.
Strontium:	crimson.
Lead:	fleeting blue.

Spot Tests

Spot tests can be made under the microscope, using a glass slide as the spotting surface and employing either inorganic or organic reagents. Here the inorganic reagents often lack the specificity of an organic compound in precipitating or color forming with a desired constituent.

Components of a particulate sample that are alkaline earth metal compounds such as lime and limestone are detectable as Brönsted bases, since they produce a red color from neutral 5% phenolphthalein when so treated under the microscope.

From a practical point of view, quite a number of analytically significant reactions occur when mixed particulate is treated with neutral as well as with acidic and basic aqueous solutions. Once the microscope has been adjusted to show a field of representative particles, a drop of water is added from the tip of a metal probe and the slide is moved to show the solid liquid interface at the edge of the drop. Migration of this interface from particle to particle by means of surface capillary action serves as a guide to relative solubility as well as to particle density. Frequently, CO_2 effervescence is observed upon the addition of

6 M hydrochloric acid to the already wet, though neutral, sample field. This ebullience from the body of the droplet indicates that soluble carbonates such as sodium carbonate are present, whereas release of carbon dioxide from single insoluble granules identifies these as the carbonate of calcium or some other alkaline earth metal. In this respect, particles of cement powder liberate CO_2 only slowly because of the encapsulated but otherwide uncombined carbonate.

Identifying Carbonate Particulate

Aerosol lime is gradually converted to calcium bicarbonate and finally to the carbonate. Laboratory tests have indicated a fine surface of CaO to be at least 50% converted to the carbonate species at the end of 48 hr at 20°C, 40% RH.[7] It is interesting to note that deeper surfaces of an air-exposed sample show much less conversion because of adsorption of CO_2 by surface interfacial granules with the formation of a carbonate skin protecting the bulk of the lime from as rapid a carbonate slaking. Thus windblown, finely divided lime (such as would tend to excape from a bag collector) with a large initial surface area that is continually renewed through turbulent aerosol abrasion would be expected to convert rapidly to the free calcium carbonate. This effect must be taken into consideration and allowances made for the age of a lime sample, which must be approached for analysis as a mixture of the carbonate, bicarbonate, and CaO. Large amounts of such a sample—for example, 1 g or more—would lend themselves readily to a differential end-point titration using a system of mixed indicators.

Analysis of Mixed Carbonate Particules

For approximate measurements utilizing the principle of this method, a 0.1 M hydrochloric acid titrant may be employed with phenolphthalein indicator, titrating to the HCO_3 end point at pH 8.4. Following the first color change, methyl orange or methyl orange modified with xylene cyanol may be added and titration resumed to MO end point. Approximate percentages of components result from the application of the following guidelines.

    1. If the number of equivalents of lime exceeds the number of equivalents of bicarbonate in solution, the resulting medium will contain only carbon and hydroxides:

$$HCO_3^- + OH(\text{excess}) \rightarrow CO_3^{2-} + OH^-(\text{excess}) + H_2O$$

Here, using the volume relationship given in Figure 25, multiequivalents of hydroxide may be obtained by multiplying the normality of the acid ($N = 0.1$) by the difference between the first and second end points as 0.1 ($V_1 - V_2$), while the equivalents of carbonate follow from 0.1 (V_2).

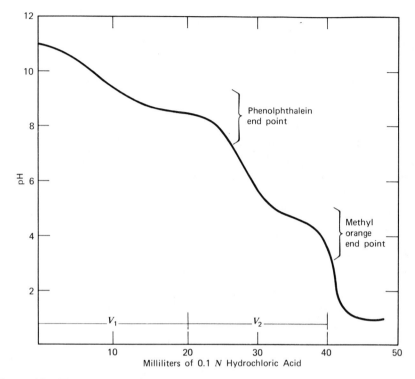

Figure 25. Titration curve for carbonate–bicarbonate particulate showing change in pH as a function of the addition of standard 0.1 N hydrochloric acid to a mixture of hydroxide, carbonate, and bicarbonate ions.

2. If the number of equivalents of bicarbonate exceed those of lime, the medium will exhibit only carbonate and bicarbonate:

$$OH^- + HCO_3^-(\text{excess}) \rightarrow HCO_3^- + H_2O + CO_3^{2-}$$

In this relationship, bicarbonate follows from 0.1 $(V_1 - V_2)$ while carbonate is determined by 0.1 (V_1).

Microchemical Technique

A very simple microchemical test relies on collection technique rather than separative chemistry to isolate individual members of an aerosol mixture. In this procedure of Crozier and Seely,[8] particles are impacted (Chapter 5) on a microscope slide that has been coated with one of a number of reagents in a suspension of gum or transparent gelatin.

Here, for example, a medium in which acidified ammonium thiocyanate has been dissolved responds to impaction of dust containing iron compounds by forming a pinkish red halo around those particles that contain iron. This is the result of interfacial diffusion reaction of the iron into the acid in thiocyanate to form the bright red $Fe(SCN)_6^{3-}$ ion. In the same test, cobalt, palladium, and platinum would interfere; however, compounds of these metals are seldom of concern in urban atmospheres.

Many such "mulls" are possible, using a wide variety of vehicle media for the inorganic (or organic) reagent. Successful sampling surfaces have been prepared with Nujol, Vaseline, gum tragacanth, gum arabic, and transparent starch. Occasionally, a sample surface that has been exposed to air currents for some time may become dry and require an additional "developing" step. The entire surface is then placed in a high-humidity chamber, such as an inverted beaker containing some source of moisture, or the sampling surface is moistened slightly to permit interfacial diffusion of reagent to particle.

Thus silver nitrate reacts with aerosol chloride to produce the white halo of AgCl precipitate, and acidic (HNO_3) ammonium molybdate gives a yellow halo around phosphate particles. Fluorides may be detected with the colored dispersion of zirconium alizarin sulfonate, where the position of a fluoride-containing particle is marked by a bleached or white spot in the pink field.

The interferences of these reactions must be kept in mind to correctly interpret the results of these diffusion plate tests, as well as the color and precipitate-forming reactions resulting from the mixing of particle samples with microdrops of reagent on the clear glass slide. Standard texts in qualitative analysis should be consulted if there is doubt about the possibility of interfering ions.[9,10] For example, silver nitrate as a reagent for chlorides will also precipitate sulfate and phosphate ions, although positive identification of such a reaction product could be made microscopically by confirming the refractive index of the product formed.

The diffusion plate test usually produces the clearest result when there is a strong source of known particles. It does serve as a rough quantitative measurement by scanning the field of total particles to determine the count of those that react with the reagent medium. In addition, the precipitate color or shade opacity of the halo is often sufficient to gauge the amount of constituent in the particles.

Separation of Heterogeneous Samples for Microchemical Analysis

Often it is desirable to separate a heterogeneous sample. An aqueous separation can be made by using a small Büchner-type funnel with a 1-cm plate and Whatman No. 1 filter paper. Rinsings can be captured in the bowl of the funnel by placing the index finger over the exit tip while filling the funnel with water to a point just above the level of the sample. Two to three such rinses, when held in contact with the sample for several minutes, are sufficient to dissolve most soluble alkali chlorides, sulfates, and oxides. The insoluble residue can then be removed from the paper and mounted on a microscope slide, or it can be viewed on the filter paper where it has been retained.

The same procedure can be used to concentrate a soluble component from a small heterogeneous sample. The aqueous rinse is then collected and evaporated to a certain amount, which is placed on a microscope slide where it is evaporated to dryness, leaving the recrystallized sample ready for viewing. Each of these evaporations should take place at or only slightly above room temperature to prevent the volatilization of water-soluble organic constituents. Since such a sample has undergone aqueous recrystallization, the chemical composition of the "natural" sample may be altered. However, this procedure is advisable when sufficient ionic species are present to disclose the source of the sample.

These microchemical tests are more appropriate in the confirmation of suspected contaminants than in wide-scale separation and identification of a number of ions or compounds. To obtain *this* type of information, other techniques such as x-ray fluorescence or diffraction, activation analyses, or emission spectroscopy must be employed. The x-ray diffraction spectrometer is one of the most useful instruments in air pollution analysis, since molecular as well as elemental information is generated.

MICROSCOPIC IDENTIFICATION BY MEANS OF CRYSTAL PROPERTIES

When the sample can be viewed by means of underlighting rather than reflected light, much is obtained through crystal birefringence. If an anisotropic substance such as silica (quartz) is viewed at an angle 45° to its plane of increasing thickness, double refraction (birefringence) occurs each time the transmitted light is reflected from a crystal plane. In large pure crystals this produces ordered interference color known as Newton's series at each periodic (90° angle) increase in the crystal's thickness.

Technique Using Birefringence and Refractive Index

If, however, such a crystalline material is ground into a fine powder, the probability of the number of small crystals assuming the proper orientation for viewing becomes very large and only total interference versus total reinforcement, or the periodicity itself, is observed. This gives rise to a color-in color-out phenomenon as the stage or platform that holds the sample is rotated through successive crystal phases from 0 to 360°. The only requirement here is a microscope with underlighting and with a rotating stage or rotating adapter mounting for the cross-polar prisms needed to view the polarized light.

A short list of substances that exhibit the property of birefringence is given in Table 2. Degree of anistropy refers to the relative refractive index (RI). If the RI is above 0.1, birefringence is high. Relative refractive indexes of less than 0.1 and less than 0.01 are considered to be medium and low, respectively. A Michel-Levy chart, included in many books on microscopy, shows the order of birefringence as a function of crystal thickness and the resulting color.

A beginner who has had some experience with the operation of a microscope can in a short time master the rather birefringent character of known samples merely by observation. This ability is best developed by first observing a series of known crystals (ground to 1.0 to 10.0 μ to produce a good field) such as silica, calcite, and lime, then gradually adding a few crystals of limestone of similar size, then adding ground "known" fly ash, synthetic fibers, and finally coal or coke dust as a blank. Such a "known sample" gradually begins to resemble a sample of average dustfall particulate.

Substances that have only one RI are of course not subject to birefringence, but they do exhibit a characteristic RI. It is seldom necessary to measure exactly the RI of an unknown crystal. With some practice one can learn to estimate the indices from the degree of contrast of the particle in a drop of liquid medium of known RI. This is based on the same principle as the disappearance of a Pyrex glass tube when immersed in toluene where the indices of these two

Table 2 Birefrigent Substances

Substance	Degree of Anisotropy
Cotton fiber	Medium
Flyash	Low
Silica (quartz)	Low
Mica	Low
Rayon (acetate)	Low
Mylar	High
Dolomite ($MgCO_3$)	High
Limestone	High
Carborundum	Medium
Cement (baked)	Medium
Sucrose	Medium

Table 3 Refractive Indexes of Common Liquids at 25°C*

Substance	RI	Substance	RI
Trifluoroacetic acid	1.283	2-Pentanol	1.416
Methanol	1.326	2-Heptanol	1.418
Ethyl ether	1.352	p-Dioxane	1.420
Ethanol	1.359	1-Heptanol	1.422
Ethyl Acetate	1.370	Capric acid	1.426
n-Hexane	1.372	Ethylene glycol	1.429
2-Propanol	1.375	Cyclohexane	1.443
2-Butanone	1.377	Carbon tetrachloride	1.459
1-Propanol	1.383	n-Butylbenzene	1.487
Heptane	1.385	p-Xylene	1.493
sec-Butyl acetate	1.387	Toluene	1.494
2-Pentanone	1.390	Benzene	1.497
Octane	1.395	Tetrachlorethylene	1.504
2-Butanol	1.395	Furfural	1.524
1-Butanol	1.397	Acetophenone	1.532
Isoamyl acetate	1.400	o-Toluidine	1.570
Cyclopentane	1.404	Aniline	1.583
2-Heptanone	1.406	Carbon disulfide	1.628
n-Decane	1.409	Diiodomethane	1.749
2-Octanone	1.414		

* From Handbook of Physics and Chemistry, 50th ed, Used by permission of CRC Press Inc. 1970.

materials are identical. The property that the RI of a liquid changes more rapidly with temperature change than does that of a solid can be used to match the index of the solid for the purpose of identification. Liquid media with a standard curve for variance of RI with temperature are commercially available. A small kit of such fluids can be assembled inexpensively from examples, as shown in Table 3. Lists of standard immersion liquids appear in, for example, *The Eastman Organic Chemical Catalog*. Such kits are also commercially available from any distributor of microscope or chemical supplies. In general, one can cover a range from 1.46 to 1.82 using only three liquids: balsam, Aroclor, and methylene iodide; this particular set of liquids covers a collective range that includes the RI's of more than 90% of the particulates most frequently encountered in air pollution analysis. Examples of substances having a single RI are given in Table 4.

It is noteworthy that substances such as sea sand, often used as an abrasive, can be differentiated from gravel or silica by shape, whereas molding sand,

Table 4 Refractive Indices of Isotropic Substances Often Found as Pollutants*

Substance	RI
Bakelite (phenol-formaldehyde)	1.48–1.65
Sodium chloride	1.544
Orlon	1.515
Polystyrene	1.59–1.67
Starch	1.51–1.54
Nylon fiber	1.52
Glass fiber	1.54
Cellulose nitrate	1.51
Vinyl chloride polymer	1.54

* From Handbook of Chemical Microscopy, vol. II, E. M. Chamot and C. W. Mason, John Wiley, New York, 1958, p. 333

which is a common industrial aerosol particulate, can be differentiated from each of the substances above by the RI of its coating of resin (usually phenolic). In a sample of sufficient size the resin can be extracted with organic solvent and recrystallized, and a confirmatory infrared spectrum determined to identify the particulate precisely.

Dispersion Staining

Dispersion staining has the advantage over other RI methods that even a nonchemist, if he has sufficient familiarity with the microscope to mount samples and observe their color, can identify unknown particulate by this form of microscopic analysis.

For example, if an aerosol of common road salt (NaCl) is expected to occur in a number of dustfall samples, its presence and a rough estimation of its relative amount (e.g., 25 or 50%) can be determined by mounting samples of the dustfall in Cargill refracting index fluid No. 1.540. The blue coloration of the salt crystals seen at 20°C in this liquid will not vary from sample to sample, nor will any other substance have exactly the same color. The reason is that the wavelength (λ) at which the temperature curve for the RI of the liquid crosses that of the solid (particulate) determines the color assumed by the particle.

Figure 26 shows the relationship between various particulates of different refractive indices (n_D) and the dispersion staining liquid Q. Where the RI of particulate crosses the curve of the liquid, the corresponding wavelength (λ = 425 mμ) is that of blue light. For particulate the crystal at 630 mμ appears orange.

Procedure for Dispersion Staining

In practice, given a sample of dust containing several types of particles, identification by means of dispersion liquids can be obtained as follows. Let us assume that at least some of the particles are recognizable by some other previously mentioned physical property, which is very often the case. Let us say that fly ash has been identified by its sintered appearance and that coke particles are apparent as coke balls having a typical shiny surface. The balance of the sample contains what appear to be tiny white to frosty white crystals of regular shape that are insoluble in water and 6 M hydrochloric acid and do not produce a red color when treated with neutral phenolphthalein. Cross-polar prisms using underlighting show no birefringence; therefore the particles are assumed to be isotropic.

A small amount of the sample is then mounted as a mixture; if greater selectivity is desired, the unknown crystals are isolated by the Pasteur method using a probe under the microscope. Either a gross sample or the separated fraction is suitable for stain mounting, where isolation merely aids in concentrating attention on the desired constituent.

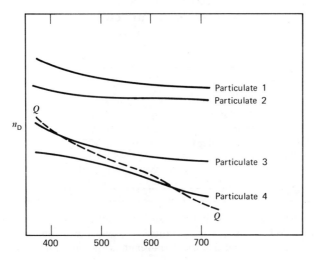

Figure 26. Plot of n_D^{25} of the dispersion liquid versus observed λ.

In practice, the unknown sample is mounted in an RI fluid (e.g., Cargille fluid) and the dispersion staining colors are recorded. If no colors are observed, the liquid selected is assumed to be beyond the RI range of the solid. If so, the following test procedure is applied to determine whether a higher or a lower RI liquid is required. Near the interfacial boundary of an unknown particle, a bright halo will be visible in sodium light. If the microscope is focused up, that is, the position of the focus is raised from the table, the halo will always move toward the medium of higher RI. Upon lowering the focus, the halo will move in the direction of the medium of lower index. It is also useful to remember that the closer the RI of the particle is to the medium, the less will be the contrast between the halo and the particle at the boundary. That is, this test can be used to determine whether the particle or the medium has the higher RI; the relative contrast shows the degree of difference.

When staining of the unknown particles is observed, the exact color is determined either from a spectral line chart, which relates optical color to λ, or by use of a monochromator.* Thus by plotting the n_D^{25} of the liquid chosen versus the λ observed on charts similar to Figure 26, showing the RI's of known compounds, the identity of the unknown follows from the point of intersection as 1, 2, 3, 4, ... The RI's of some common air pollutants are shown in Table 5, and the extreme indices of several anisotropic compounds are given in Table 6.

Table 5 Refractive Indices of Some Common Isotropic Particulates

Substance	RI
NaCl	1.544
Cr_2O_3	2.5
Fe_3O_4 (magnetite)	2.42
CaO (lime)	1.838
CaS (sulfur slag)	2.137
NH_4Cl	1.640
MgO	1.736

---

* Monochromator adaptors can be purchased from all well-known manufacturers of optical supplies. When using a monochromator together with an annular stop, the desired λ is the wavelength noted at which transmittance occurs with simultaneous disappearance of the particle.

Table 6 Refractive Indices of Common Anisotropic Particulates[a]*

Substance	r_α	r_ϵ
Fe_2O_3 (hematite)	3.22	2.94
SiC (carborundum)	2.647	2.693
TiO_2 (titania)	2.605	2.901
FeO0OH (rust)	2.260	2.515
ZnO (zincite)	2.013	2.029
$CaCO_3$ (limestone)	1.658	1.550
SiO_2 (quartz)	1.544	1.553
$CaSO_4 02H_2O$ (gypsum)	1.519	1.531

[a] Only extreme indices are given.
* From Handbook of Tables for Applied Engineering Science, Used by permission of CRC Press Inc. 1970.

Some Precautions in Dispersion Staining

When the temperature in the laboratory varies from 25°C, an appropriate correction factor must be applied, as explained in the directions for using Cargille liquids. This variation of density with temperature extends the effective range of the liquid but works to the disadvantage of the operator who might expend long periods of time in particle inspection while the sample is mounted and immersed. Illumination light (Na or yellow filter) transmitted directly through the sample is absorbed efficiently by glass slides, which act as insulators for short periods of viewing. If, however, more than 4 to 5 min are required for inspection of a particularly complicated sample, the light should be periodically extinguished to equilibrate the room temperature of the sample and its medium.

It also is well to remember that dispersion staining is most easily applicable to isotropic crystals, whereas birefringence characteristics are best applied to anisotropic substances. This does not mean that identification of anisotropic particles is not possible by the dispersion technique. However, the determination of the two most extreme indices is more difficult and involves rotation of the polarizer to find the λ_1 and λ_2, which may often require separate mountings in two different dispersion liquids for comparison of the two n_D^{25} values of the anisotrope. Such an approach is more suitable for confirming the presence of a suspected contaminant particle than for the identification of a totally unknown constituent, the difference between the two being only the amount of time involved. As a rule of thumb, only glasses, a few plastics, and compounds with a cubic crystal system are isotropic.

It is well to note that the optimum particle diameter range in the identification of particles by dispersion measurement is on the order of 0.5 to 5 μ.

ASSEMBLING A LIBRARY OF KNOWN PARTICULATES FOR MICROSCOPIC REFERENCE

Many particles are easily recognizable once the eye has become familiar, through the study of knowns, with their size, shape, color, and such other physical properties as smoothness, opacity, spiked appearance, granular surface, cell-like structure, and metallic stress (coloration inside metal turnings). Photomicrographs are also helpful. An even better practice is to file small vials of knowns rather than photomicrographs, since a known provides a larger population of particles than does a single microscope field or even the two or three fields of stored photographs or transparencies.

Such knowns as fly ash from a particular power plant can be collected from the stack precipitator and filed together with known slag, cement powder, white metal emission, and others. A substantial amount of knowns can be stored in such a library, which may be useful for chemical as well as microscopic analysis should a comparison be desired at some later date. In addition, a residual contaminant of the large and otherwise "pure" sample can be concentrated by the scanning and selective probing of a 1-oz sample to provide trace contaminant evidence which, should it have been overlooked previously, may help to relate a new aerosol residue directly to its source.

Small amounts of glass fiber, included in a sample of lime taken from a bag house, for example, could be overlooked until much later when their recognition in the bulk sample bottle might lead to further inspection of dustfall collectors in the vicinity of the lime plant; such secondary evidence might separate the emission of a kilning operation that used fiberglass bags from an identical emission that had been collected in Teflon or nylon bags.

Also, trace contaminants such as carborundum wheel grindings in copper particulate or catalyst residue in petroleum coke could lead to identification of source, if the chemist or technician is alert to such clues when comparing "old knowns" to recent samples of emission.

X-RAY DIFFRACTION IDENTIFICATION OF PARTICULATE

The most recent and by far one of the most useful instruments in the field of air pollution analysis is the x-ray diffractometer with its spawned microextensions: the electron microscope, scanning electron microscope, and the electron microprobe. For this reason their use and general theory of operation are discussed in some detail.

With the recent growth in publication of standard indices handbooks,[11,12] the labor usually associated with interpreting both diffraction spectrograms and powder patterns had diminished markedly. As a result, such analysis may now be applied almost routinely to the determination of pollutant composition in tracking down a source of the contaminant or in elucidating the mechanism of aerosol interaction.

The most important application of this instrumental approach is the gaining of true molecular knowledge from the particle matrix. Whereas emission and atomic absorption spectroscopy yield only the specific ionic or elemental composition of the pollutant mixture—that is, percent sodium or percent sulfate—x-ray diffraction lines represent energy interaction transmitted by the entire compound or crystal unit whether it be $Na_2CO_3 \cdot 10H_2O$ or o-dichlorobenzene. Thus together with the chemically more informative technique already discussed, diffraction yields the correct stoichiometric form of aerosol compounds, which when mixed might show chemical analysis of Ca, Na, Ti, Cl, S, and O, but when reconstructed by means of a diffraction spectrograph could conform to CaO, TiO_2, $NaCl$, and $CaSO_4$.

It is the feeling of this author that the average chemist who is willing to devote some time to learning the general position of the classes of diffraction frequencies and to developing an eye for the perception of slight shoulders in chart traces—the distinguishing of noise from signal at critical frequencies and the preparation of reproducible samples in spite of occasionally tedious grinding procedures—will find in this approach a very analytically rewarding means of sample determination. The mathematical demands are not so formidable as one might suppose, requiring only a knowledge of the general techniques of chart reading and peak integral area calculations such as are used to quantitate the chart readouts from gas chromatography or infrared spectroscopy.

Although the discussion in this section is directed toward actual particle identification, some material is presented on the theory of operations; following the assumption that to make the best possible use of these methods and to learn other similar methods, we must first at least briefly explore the phenomenon of diffractive crystal structure.

Theory and Operation

To understand the phenomenon of x-ray diffraction, we should first examine three concepts: (1) the nature of x-rays, (2) the nature of crystal structure, and (3) the meaning of diffraction.

The series of waves that occupy the region of the electromagnetic spectrum between UV light and gamma radiation are called x-rays. These waves have a λ range between 0.1 and 100 Å and are usually produced by focusing a beam of electrons from a heated cathode onto a metal target (Figure 27).

Since not all of these electrons stop at a single energy distance into the target, a small *spectrum* of waves is produced. This spectrum known as white radiation has an intensity peak centered rather sharply in one or perhaps two specific wavelengths (Figure 28).

The minimum wavelength at the threshold of this spectrum is determined by the voltage energy that produces the stream of electrons, while the λ position of the intensity maximum is determined solely by the metal chosen as target material. Note that the relationship of characteristic x-ray λ_α to the various elements is expressed through $c = \lambda v$ in the well-known law of Moseley, where $N = k/v$. Here N is the atomic number of the element and k is a constant representing the direct straight-line variance of atomic number with increasing x-ray frequency (v).

For reasons that will become clearer in the discussion of selectivity, the wavelength used to irradiate the sample must be as monochromatic as possible. Either the K_α or the K_β line can be used; however, it is more practical to filter the shorter wavelength, hence higher energy, K_β line with a selective filter, which is fabricated from a metal that will absorb the K_β radiation without causing scattering or producing fluorescence. Such a metal usually immediately precedes the target metal in the periodic chart. Thus a nickel filter passes specific K_α radiation from copper and a manganese (or MnO_2) filter passes K_α from iron where these two particular energies are among the most common and together the most useful in application to air pollution analysis. These comparatively low energy λ's may be excited by 35 to 40 kV electron beams and may, therefore, be produced by more economical source equipment than that required for the higher energy x-ray fluorescence function as well.

In the study of crystals by x-ray diffractometry, the analyst discovers repeated evidence that the entropy of pure crystalline solids is unity. The

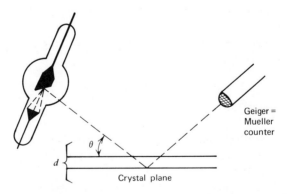

Figure 27. Production of x-rays by cathode irradiation of a metal target.

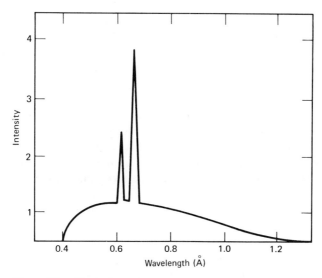

Figure 28. Typical x-ray energy spectrum.

perfect organization of these structural units gives rise to reproducible crystalline spectra when a beam of monocromatic x-radiation of the same magnitude as the spacings between crystal plans is scattered from the internal lattice surfaces of a crystal.

The original Laue method of diffraction is perhaps the purest classical method because it is the theoretical "sample crystal" that is irradiated. This arrangement provides the three-dimensional scattering required by the Bragg relationship of scattering angle with energy:

$$n\lambda = (a)(\cos \alpha - \cos \alpha_0)$$
$$k\lambda = (b)(\cos \beta - \cos \beta_0)$$
$$x\lambda = (c)(\cos \gamma - \cos \gamma_0)$$

Solving these equations simultaneously produces the familiar Bragg equation

$$\lambda n = 2d \sin \theta$$

where λ is the wavelength of the x-radiation, d is the vertical spacing between the crystal planes, and n is an integer representing the order (first to second, second to third, etc.) of the crystal planes. Here θ is the angle of the diffracted beam of x-rays, and it is this variable that is of most use to the qualitative analyst who approaches diffraction as a means of identifying "unknown" crystalline solids.

An alternative to the Laue method, which requires the mounting of a single crystal of material, is the powder diffraction method, which necessitates only grinding of the sample to produce a thin crystalline powder whose average diameter is on the order of 1 to 10 μ. The random orientation of these crystals yields good statistical probability that a number of these particles will be oriented with alignment of Miller indices to satisfy the Bragg equations and produce a pattern closely resembling three-dimensional Bragg symmetrics that have been rotated through 180 planar degrees.

A photographic negative of this Debye powder pattern resembles Figure 29. The pattern is produced by mounting the thin film of finely ground particles either directly in the transmission path or along an angle of reflection to the beam of monochromatic x-radiation. Since the λ of the x-radiation is known and since it can be calculated from the position of a given line when the exact sample to film distance (this is built into the spectrometer) is known, the only remaining variable is d, the interplanar distance, which is a function of the atomic arrangement of a pure element or the molecular arrangement of a compound.

Each substance has a unique order and number of such spacings, which can be used as a total fingerprint to confirm the presence of the substance. Not only can a substance be so identified, but quantitation can be achieved by using homogenized powder solution standards. In addition, different crystal forms of the same substance such as α- or β-quartz can be separated and each allotrope quantitated, provided that suitable pure standards are available.

The information may become important in solving a problem in which the "history" of the sample must be determined. Let us say that a sample of aerosol suspected of containing calcium carbonate has been collected at some recent time and is being compared to other samples obtained from dustfall containers or perhaps eaves and window ledges in the same area. In acquainting ourselves with the geochemistry of calcium carbonate, we would find that the calcite form is thermodynamically stable and the aragonite form is kinetically stable. We would expect the calcite to be present in the residual material collected but would not be surprised to find some aragonite, in aerosol especially, if the emission source were lime or cement kilning and the carbonate present were formed only recently from aerosol reaction of lime with CO_2.

While a given crystalline solid may have a number of interplanar distances, only two or three are required to identify the solid. These values or d spacings have been tabulated in a form arranged by Hanawalt, Rinn, and Frevel.[13] The following section describes their use.

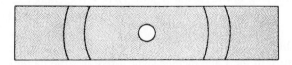

Figure 29. Typical Debye Powder Pattern.

Illustrative Example of Particulate Analysis

Let us assume that a sample of particulate has been delivered to the laboratory with the request to determine the percentages of gypsum, limestone, and lime in as short a time as possible. The sample is grayish brown and particles are on the order of 20 to 50 mesh (coarse sand).

Information Achievable by Chemical Analysis Versus X-Ray Analysis

A chemical analysis would probably require determination of the free sulfate by precipitation in acid solution (to prevent coprecipitation of carbonate). Subsequent analysis of the cement as silicate would separate cement from noncement calcium but would require a rather tedious separation of cement silicate.

If it could be shown that no other contaminants were present, the cement could be assumed to be insoluble in dilute hydrochloric acid with the balance of the aqueous solution subject to the molecular equivalents of $MgSO_4$ and CaO based on a determination of the calcium magnesium equivalents by the Winkler method[14] or, more conveniently, by atomic absorption spectrophotometry.

Total carbonate could be determined by using 6 M hydrochloric acid to evolve CO_2, which is then carried through a calcium chloride drying tube into a weighed Ascarite tower. Any residual iron (from cement present) could then be titrated directly in this solution using standard potassium dichromate once the separation of CO_2 was complete.

Such an analysis would produce the following information:

1. Total silicate.
2. Total calcium.
3. Total magnesium.
4. Total carbonate.
5. Total sulfate.

A differential indicator titration would yield carbonate, bicarbonate, and total free base; however, assignments of calcium ion to the correct equivalent of carbonate and the determination of cement silica and silicates would furnish the analyst with, at best, only two significant figures.

The same analysis could be carried out by x-ray diffractometry in a fraction of the time with thorough confidence in the resulting percentages of *molecular* mineral components.

Procedure for X-Ray Diffraction Analysis

The sample would first be ground. Hand grinding would probably require 10 to 30 min to reduce the particles to an average of 1 μ diameter. An alternative method is the use of an automatic mortar and pestle, or vibrator, suitable for powdering minerals; here the grinding time is reduced to 1 to 5 min. Some research has been done on the relationship of grinding time to actual diffraction peak intensity, since this aspect becomes important in quantitating diffraction spectra.[15] A rule of thumb when initial particle sizes of standards and samples are roughly comparable is to grind sample and standard for equal times and to add 5 min of equivalent hand grinding for each 50 μ of estimated average diameter difference between the coarser and finer materials. Most aerosol samples collected in urban or industrial areas will characteristically contain a wide variety of initial particle sizes. Moreover, frangibility and hardness of particles within the same sample may vary considerably. Such problems are often resolved by manually separating the particles that during the first few minutes of hand grinding appear larger and harder. This rather refractory material can be ground separately and then readmixed with the sample. In addition, it is often helpful to employ an internal standard such as calcium carbonate or kaolin when quantitating samples to minimize the effect of grinding time on the intensity of diffraction peaks. The ratio of the integral area of the main 2θ peak of the internal standard to one or more selected sample peaks is then used as the parameter by which to compare relative peak heights. Figure 30 shows the relationship of actual peak height to grinding time for silica sample.

Following sufficient grinding, the sample is mounted on the planchette with a spatula, as illustrated in Figure 31.

Care must be exercised that the sample well or frosted surface in the planchette is uniformly covered with powder and that firm strokes of the spatula result in a well-packed surface that will not detach during mounting on the x-ray table. A particularly dry or powdery sample may be mixed with a

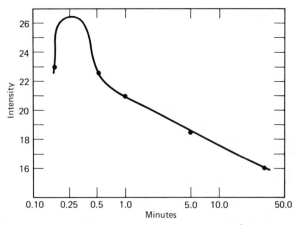

Figure 30. Intensity of diffraction energy of α-quartz 4.26 Å d line as a function of powdering time at 1×10^3 cps. (from reference 7)

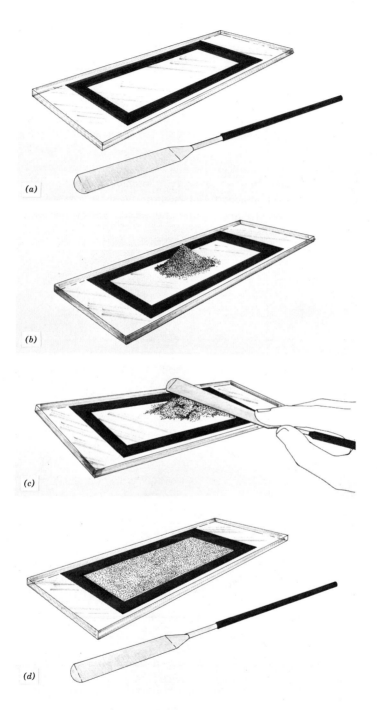

Figure 31. (a)–(d) Preparation of planchettes by spreading powder as shown.

weighed amount of kaolin or powdered glass[7] to the extent of 25 to 50% of the weight of the sample to cause sufficient aggregation for mounting. Standards that contain the same proportion of kaolin will yield equally intense diffraction lines.

The planchette mounting that follows places the planchette, containing a well-packed film of powder, into the clip arrangement, which secures the powder at a fixed angle in relation to the path of the x-radiation. A sample that is not level with the surface of the planchette but has irregular "hills and valleys" may cause a shift in θ, the diffraction angle, which could result in misinterpreting the identity of one or more components.

Further directions for handling and mounting powder samples are given in manufacturers' instruction manuals, and any questions about procedure should be directed to such sources of information.

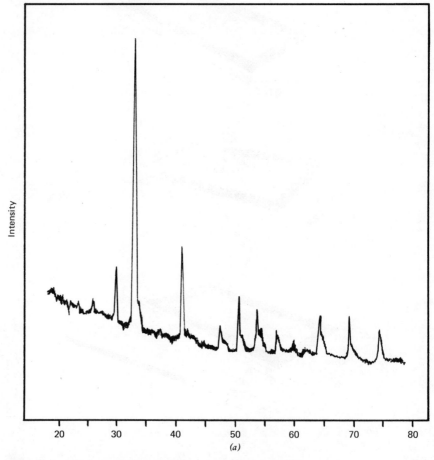

Figure 32. (a) X-Ray diffraction spectrum of pure silica, SiO_2.

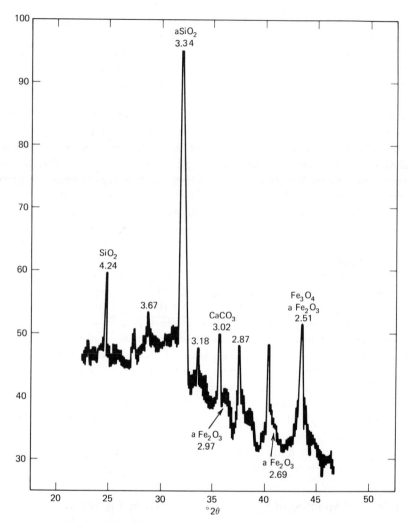

Figure 32. (b) X-Ray diffraction spectrum of a typical dustfall sample using Fe α-source of energy (from reference 7).

For analytical work, scanning angles usually begin at $\theta = 10$ to $15°$. Lower angles often produce only reflection scattering, and angles close to zero result in broadcasted x-radiation, which may overload and damage the goniometer.

Interpretation of Spectra

The spectra in Figure 32 illustrate the diffraction lines of pure standards. This gives an insight into the total structure of peaks, allowing interpretation of fine structure and enabling the separation of overlapping lines that may appear in an actual sample of aerosol or dustfall particulate.

Spectra such as these result from attaching a recorder to the voltage output of the Geiger Mueller counter, which acts as x-ray goniometer. A more exact method of determining diffraction line intensities is achieved by means of a scalar counting of the Geiger pulses. Here the actual total pulse height or an attenuated portion of this signal is collected arithmetically by decade. Integration of peak height is thus achieved without graphical analysis.

In the previous example, the spectra of known standards were merely compared with those resulting from the sample to arrive at sample identification where only a given number of industrial contaminants are expected. However, incident energy is absorbed by combinations of metallic trace elements acting as filters to the x-ray path. When a question arises about the possible absorption by a compound of such a metal, the sample may be "spiked" with a known amount of the suspected compound—such as MnO_2, Fe_2O_3, or Cr_2O_3—to determine the effect, if any, and the extent of the absorption.

When "knowns" are not available to confirm the identity of suspected components, the table of standard d spacings (Hanawalt table) serves as a good means of sample identification, provided that there are only a few components and that they are more than 5 to 10% of sample weight.

Such an assignment of diffraction lines might be performed as follows (Figure 33). In an actual diffraction pattern of an "unknown aerosol" mixture we find a very common arrangement of peak intensities. Reading from left to right, we scan through increasing angle 2θ. After the θ location of all major and minor peaks has been noted, d spacings corresponding to each peak should be arranged as illustrated in Table 7, by means of the Hanawalt or Smith reference works.[16]

In these references, a table of decreasing d values relates major peaks for any crystal listed to two other minor peaks for the same crystal. Consequently, one

Table 7 d Spacing of Chart Peaks Shown in Figure 33

Peak	2θ	d	Compound	Relative Intensity (%) to Strongest Line Peak
a	26.3	4.26	α-Quartz	35
b	33.7	3.34	α-Quartz*	100
c	37.1	3.04	$CaCO_3$*	100
d	46.3	2.46	α-Quartz	7
e	50.0	2.29	$CaCO_3$	18
f	54.9	2.10	$CaCO_3$	18
g	64.4	1.82	α-Quartz	17

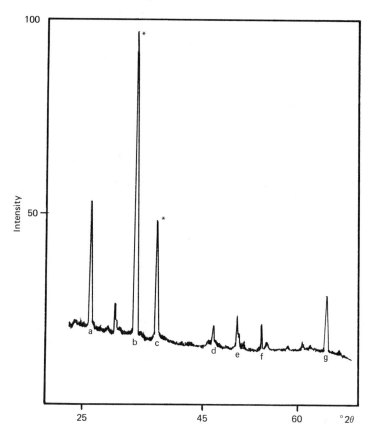

Figure 33. Illustrative example: x-ray diffraction spectrum of an "unknown aerosol" mixture.

needs only to determine the apparent major peaks in an unknown pattern and then scan the compounds that have major peaks at or around this value. The field of choices may be narrowed by successively comparing minor peaks. Necessary d spacing information required for the use of these charts is found by means of the Bragg equation, solving for d, and using the λ of the x-radiation from the element used as the energy source.

For the convenience of the reader, the K_α for a number of common metallic targets are shown in Table 8.

The reader should keep in mind, however, that many instrument manufacturers provide a table of d spacings indexed to the 2θ observed during use of targets supplied with their instrument. These tables are very helpful in quickly assigning d spacings to chart peaks before employing a reference index for identification.

A scan of the *Index to the X-Ray Powder Data File*[16] shows quickly that the

Table 8 Radiation Wavelength of Common X-Ray Target Materials

Source	K_α(nm)	K_β(nm)
Copper	15.40510	13.92170
Iron	19.35970	17.56530
Chrome	22.89620	20.84800
Tungsten	2.08992	1.84363
Nickel	16.57840	15.00100

peak c at $2\theta = 37.1°*$ is the maximum intensity peak for calcium carbonate, which is confirmed by matching peaks e, 50.0°, and f, 54.9°. The other major peak in this example pattern occurs at b, 33.7°*, to indicate silicon dioxide as α-quartz whose minor peaks at a, 26.2°, g, 64.4°, and d, 46.3° confirm the identification (Table 7).

Quantitation of Diffraction Spectra

Quantitation of this unknown could now be achieved by diluting the unknown 1:1, 1:2, and so on with powder dilutions of limestone and of silica (α-quartz) in kaolin and using the major kaolin peak at $2\theta = 31°$ to establish a ratio. This "ratio" method can be avoided by a recently developed approach using powdered glass as an amorphous matrix[7] where peak areas can be plotted directly.

A chart margin integrator is useful in obtaining the integral of these areas; however, in the absence of such an accessory, a weight of the peak that may be measured by cutting the actual peak from the chart paper can be compared with a weighed square centimeter of paper to determine square centimeters of peak area.

Quite often an identification such as that performed above is rendered difficult because of overlapping or closely spaced major or minor peaks. Since the diffraction pattern has very little to do with the chemistry of the crystal, it is frequently simple to determine whether, for example, a peak at $2\theta = 35.0°$, which could belong to sodium iodide or cadmium borate, is conclusively one or the other by a simple chemical test or flame color confirming the presence of one or more of the suspected ions.

Determination of Partially Crystalline Substances

An example of the application of x-ray diffractometry to an analytical problem not given to solution by wet chemistry is the separation of coal and coke. As already mentioned, this separation can be undertaken by specific reflectance, in which the sample is dissolved in epoxy resin and allowed to set to the shape of a small cylinder. A slice of this material can then be polished and mounted in a narrow aperture microscope. Individual particles are illuminated with top lighting, and the total reflectance light of a population of not less than 2000 particles is used to estimate the proportion of coal to coke.

A much less tedious alternate method follows from x-ray diffractometry provided that the coke contains some amount of graphite. Ordinarily "high temperature" coke is produced at temperatures exceeding 2000°F, at which sufficient graphite is formed to yield detectable graphite lines on x-ray powder diffraction. Figure 34 shows a typical sample of high-temperature coke in which the graphite line at $2\theta = 33.3°$. The second and third synthetic graphite lines at 70.1 and 56.6° represent only about 3% of the amplitude of the strongest line and do not always appear.

Here the x-ray quantitation of coke particulate from a dust sample can be

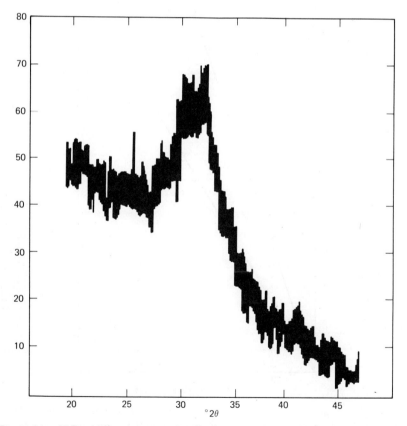

Figure 34. X-Ray diffraction spectrum of coke particulate using Fe α-source of energy (from reference 7).

achieved by mixing the sample with powdered glass (amorphous matrix) and plotting the weight percentages of coke according to a standard curve, as shown in Figure 35. Similar quantitation of airborne industrial slag dust is shown in Figure 36.

ELECTRON MICROSCOPIC IDENTIFICATION OF PARTICULATES

The x-ray diffractometer is a very useful analytical instrument for particles of 5 to 10 μ; its usefulness diminishes markedly if the particles to be analyzed are smaller than 1 μ. Consequently, particles in the range of 0.01 to 0.06 μ, a common range for open hearth furnace and automotive emission, are lost to the

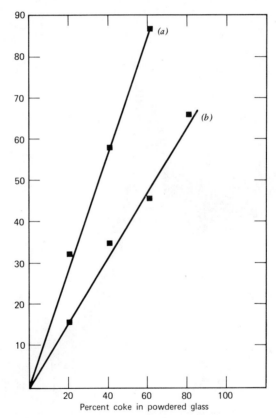

Figure 35. Reproducibility of intensity-concentration relationship for coke in powdered glass diluent over 1X (a) and 2X (b) counting attenuation (from reference 7).

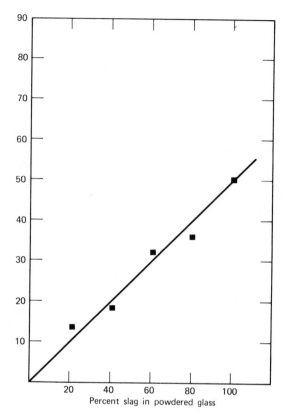

Figure 36. Reproducibility of intensity-concentration relationship for industrial slag dust in powdered glass diluent (from reference 7).

long-wave x-rays and require a shorter wave such as electron energy for diffraction.

The instrument of choice for such particles is the electron microscope, whose image can best be described as a reflection or a shadow of the actual particle where the distinctness of shape is a function of scattering effectiveness and crystal order of diffraction. While the former describes the face of the particle that is scattering, the latter provides a coefficient of crystal d spacing that, like the d spacing indices in x-ray diffraction, provides a ready means of qualitative identification.

One main feature of an electron beam is its very short wave, which is also its chief drawback in terms of the materials that lend themselves to this technique. The actual penetrating power of high-energy electrons precludes the use of this approach in the structural analysis of the relatively soft organic crystals and even a rather wide variety of inorganic crystals. Generally, the technique used to minimize this electron penetration problem invokes vacuum deposition of a metal, such as platinum or tungsten, onto the surface of the sample crystals. The "hard" metal then acts as an energy scattering mirror to produce a true image of the otherwise "soft" crystal.

Where carbon black, a common air contaminant originating from rubber tires, occurs in the 0.01 to 0.1 μ range, such techniques are not only useful for identification, but are selectively opportune in that they reveal several kinds of structures for various carbon blacks depending on conditions of formation. In addition, the image shown by the dark field electron scanning microscope is unique to carbon black, showing small rings of carbon atoms usually containing eight atoms per ring.

Theory and Operation

The scanning electron microscope includes an electron focusing system, which produces a probing beam of electrons about 5 Å in diameter. The crystal or powder of unknown material is placed at the focus and the sample scanned. Electrons are gathered by a bowl-shaped detector that is placed below the sample in such a way that electrons scattered at angles of 50 to 100 mrad are counted.

In electron microscopy, where N electrons are present in the electron probe, some electrons N_s will be scattered—some of them, N_i, inelastically and some, N_e, elastically. Thus $N_s = N_i + N_e$. In addition, a certain number of electrons N_v will pass in a vector directly through the sample:

$$N_v = N - N_s.$$

The ratio N_e/N_i is then employed through the relationship

$$\frac{N_e}{N_i} = \frac{kZ}{19}$$

to determine the atomic number Z of the scattering nucleus.[17]

Elastically scattered electrons serve to scan a reflected image of the crystal, and some extremely sharp electron shadow "pictures" have been produced by this technique. A number of air contaminant related studies are presently being pursued to determine the transport aerodynamics of aerosols through calculations based on the actual shape of the single particle as revealed by electron scanning microscopy.

SUMMARY OF METHODS USED TO CHARACTERIZE PARTICULATES

As a basic analytical device for the study of aerosol particulates, the microscope has many advantages over other laboratory instruments discussed in the preceding sections. X-Ray powder diffraction on less than a few milligrams of sample is at best difficult, yet this is occasionally the sample size left as evidence of some episode of pollution. In such cases a template or notch may be fashioned by the analyst and the sample packed into this small surface in the center of an ordinary planchette or at the point in the area of the planchette that is the known focal point of the x-ray beam. Here a planchette filled with a light-sensitive compound such as AgBr may be aligned with the sample slit by using pencil markings to show the position of the planchette in the sample housing. A few minute's exposure to the x-radiation should produce sufficient darkening to autoradiograph the focal point of the x-ray beam, allowing a good positioning of the template with its microsample.

Not all of the disadvantages of x-ray diffraction are so easily circumvented. It should be kept in mind that many of the most common air contaminants are virtually the same sort of material. Thus a common mixture of silica, road dust, glass wool fibers, utility plant fly ash, and diatomaceous earth, when ground together, produce approximately the same diffraction pattern—that is, that of SiO_2—with only background variation to suggest that several different forms of this material are present. The microscope, however, permits a visual, commonsense analysis of what the x-ray diffractometer sees. Fly ash is thus visually distinguishable from glass wool, and ground glass, amorphous to the diffractometer, can be easily recognized microscopically by its refractive index and thus separated from silica.

Assuming an average density, particles on the order of 1 to 5 μ usually weigh about 1×10^{-10} g. Since this particle size is within the range of the common optical microscopy of 30 to 400×, a single particle (whose sample weight is approximately 0.1 ng) can be at least qualitatively identified. Furthermore, if the density of the particle can be estimated once its identity is established, it is a simple matter to calculate its weight using Lenz' law to obtain its volume:

$$\frac{d_i}{d_o} = \frac{\text{image size}}{\text{object size}} = \text{order of magnification}$$

Here d_i and d_o are the image and object distances, respectively, and can be obtained from focal length data supplied with the microscope. Image size can be estimated using an image splitting attachment; or a standard particle of known size or a micrometer may be placed in the field along with the unknown particle and an "object size" determination made directly.

Very often resolution of single particles can be improved by recrystallization of the sample referred to earlier. This may be accomplished by a drop of water or a polar solvent such as tetralin, tricresyl phosphate, or methanol on the microscope slide containing the sample. To avoid poor separation of crystals, the sample and slide should then be placed in a covered container in a cool place to prevent rapid evaporation of the solvent.

The many techniques of both optical and electron microscopy discussed in this chapter might lead the reader to conclude that microscopic analysis requires a great deal of technical background. This is simply not the case. It is true that much time and effort must be expended by the air pollution analyst in becoming familiar with the appearance of various particulates to develop confidence in his ability to identify unknown materials. A similar effort is required for the chemist-microscopist to become facile in his determinations of crystallographic data, such as refractive indices. However, simplified procedures are available in manuals supplied with the microscope, which, together with some of the general works in laboratory microscopy,[18,19] should serve the analyst in developing procedural approaches to particle identification.

Probably the greatest single aid in the area of visual recognition of particles is *The Particle Atlas*,[20] which contains several hundred photomicrographs of representative air pollutant particulates. Such contaminants as spores, diatoms, starches, coal soot, oil soot, various fly ashes, and many other dust components are pictured as they appear under the lens of the microscope. This reference work is especially helpful when the analyst encounters a sample whose particles are totally unlike any he has seen before. It is also wise at such times for the analyst either to prepare his own photomicrograph of the sample or to fix a portion of the sample to the microscope slide for future reference. Such an index or library of actual samples or typical photomicrographs is especially necessary in large urban air pollution laboratories where a wide variety of particles are encountered. Moreover, these slides are valuable in following the progress of air pollution abatement if samples are collected routinely from the same general areas of the city over a period of years.

A DISCUSSION OF AMBIENT AIR QUALITY CRITERIA INVOLVING PARTICULATE MATTER

Over the past few years, dating back to the Clean Air Act of 1967, the United States Department of Health, Education, and Welfare, and the Environmental Protection Agency have been engaged in assembling information to relate the present scientific knowledge regarding the various adverse effects of particulate air contaminants on man and his environment to the actual concentrations of such air pollutants.

In these documents, the following properties of particulate pollutants have been considered in arriving at ambient air quality criteria[21] for their concentration:

Concentration in air.
Chemical composition.
Mineralogical structure.
Adsorbed gases.

Coexisting pollutants.
Physical state of pollutant.
Rate of transfer to receptor domain.

These are the properties that have been discussed in this chapter, with the exception of procedures for the degassing of particulate samples where such procedures are not as yet used routinely by either National Air Pollution Control Office of the Environmental Protection Agency or local air pollution control agencies.

In terms of the definitions set forth in air quality criteria for particulate matter, it is essential that analytical methods be sufficiently sensitive to determine the chemical composition, if need be, of 75.0 μg of sampled dust to represent 1 m^3 of air, which is the present air quality standard for aerosol-suspended particulates.[22] These particulates are by definition aerosols that will be retained by a high-volume sampler operating at a flow of 40 to 60 ft^3/min under conditions that will collect particles of 10.0 μ to a lower limit of 0.1 μ. Temperature and atmospheric pressure are specified by the same definition at 20°C and 760 mm Hg, respectively.

This value of 75 μg/m^3 refers to an annual geometric mean and, in conjunction with a 24-h maximum of particulate (260 μg/m^3), constitutes the air quality that is designated as "primary" standard. "Primary" in this connection refers to those standards that are immediately human-health related, whereas secondary standards established for the same contaminant pertain to concentrations considered safe for plants, material surfaces, and other possessions of man.

These standards are, to the best estimates of the United States Public Health Service, levels of contaminant at which a factor of safety ensures the absence of detrimental effects on the man-related environment.

Note that the definition of particulate that includes a diameter range "larger than small molecules, but smaller than 500 μ" includes a particle range from 100 to 500 μ which lies outside the collecting efficiency of the high-volume sampler. This is not an oversight. Particles in this region are subject to settling forces under the influence of gravitational attraction, while particles in the 0.10 to 10 μ range possess such small masses that these behave more like gas molecules than particles and remain suspended in air. It is these that we refer to as *suspended* particulate and whose concentration sets the aforementioned air quality standards.

REFERENCES

1. H. E. Rose, *J. Appl. Chem.*, **7**, 244 (1957).
2. F. Feigl, *The Chemistry of Specific Selective and Sensitive Reactions,* Academic Press, New York, 1949.

3. F. Feigl and L. Baumfeld, *Anal. Chem. Acta,* **3,** 83 (1949).
4. F. Feigl, P. Krumholy, and E. Rajmann, *Mikro-Chem.,* **9,** 395 (1931).
5. F. J. Welcher, *Organic Analytical Reagents,* Van Nostrand, New York, 1947–52.
6. H. B. Elkins, *The Chemistry of Industrial Toxicology,* Wiley, New York, 1963.
7. P. O. Warner, L. E. Saad, and J. O. Jackson, "Identification and Quantitative Analysis of Particulate Air Contaminants by X-Ray Diffraction Spectrometry," *J. Air Poll. Control Assoc.,* **22,** 887–890 (1972).
8. W. D. Crozier and B. K. Seely, "Technical Report," *New Mexico School of Mines, Research, and Development Division,* **1,** 20 (1949).
9. F. Feigl, *Laboratory Manual for Spot Tests,* R. E. Qesper, transl. Academic Press, New York, 1943.
10. E. M. Chamot and C. W. Mason, *Handbook of Chemical Microscopy,* Vol. II, 3rd ed., Wiley, New York, 1959.
11. C. H. Macgillavry and G. D. Rieck, eds., *International Tables for X-Ray Crystallography,* Vol. III, Kynoch Press, Birmingham, England, 1962.
12. N. F. Henry, H. Lipson, and W. A. Wooseter, *The Interpretation of X-Ray Diffraction Photographs,* Macmillan, New York, 1960.
13. J. D. Hannawalt, H. W. Rinn, and L. K. Frevel, "Chemical Analysis by X-Ray Diffraction, Classification and Use of X-Ray Diffraction Patterns," *Ind. Eng. Chem., Anal. Ed.,* **10,** 457–511 (1938).
14. H. H. Willard and N. H. Furman, *Elementary Quantitative Analysis,* Van Nostrand, New York, 1940, p. 339.
15. N. A. Talvetie and H. W. Brewer, "X-Ray Diffraction Analysis of Industrial Dust," *AIHAJ,* **23,** 214 (1962).
16. J. V. Smith, ed., *Index to the X-Ray Powder Data File,* American Society for Testing Materials, ASTM Publ. 48-1, Philadelphia, Pa., 1960.
17. A. V. Crewe, J. Wall, and J. Langmore, "Visibility of Single Atoms," *Science,* **168,** 1338 (1970).
18. T. M. Hardy, "Microscopic Methods Used in Identifying Commercial Fibers," U.S. National Bureau of Standards, Circ. 423, 1939.
19. A. H. Bennett, H. Jupnick, H. Oslerberg, and O. W. Richards, *Phase Microscopy, Principles, and Application,* Wiley, New York, 1951.
20. W. C. McCrone, R. G. Drafty, and J. G. Delly, *The Particle Atlas,* Ann Arbor Science Publishers, Ann Arbor, Mich., 1967.
21. *Air Quality Criteria for Particulate Matter,* U.S. Department of HEW, PHS, National Air Pollution Control Administration, Washington, D.C., PHS Publ. No. AP-49, 1969.
22. "National Primary and Secondary Ambient Air Quality Standards," *Fed. Regist.,* **36** (1) (January 30, 1971).

2

SOURCES AND MEASUREMENT OF ORGANIC AIR CONTAMINANTS

INTRODUCTION

The publication of Federal Ambient Air Quality Standards in April 1970 has turned ambient air monitoring away from measurements that produce only total hydrocarbon toward selective measurement of the nonmethane portion of ambient hydrocarbons. The reason for this shift in emphasis is twofold. First, the nonmethane unsaturated hydrocarbons are very significant in the formation of photochemical smog, as shown in the aerosol reactions

$$NO_2 \xrightarrow[3.4 \times 10^3 \text{ Å}]{UV} NO + O \tag{1}$$

$$O + O_2 + \text{HC intermediate} = O_3 + \text{HC intermediate}^1 \tag{2}$$

$$O_3 + NO = NO_2 + O_2 \tag{3}$$

$$2SO_2 + 3O_2 = 2SO_3 \tag{4}$$

$$\text{Additional HC intermediate} + O_2 = HC - (OO) + NO_2 = RCO(OO)NO_2 \tag{5}$$

Further information on the mechanisms of these interactions can be found in the introductory work of Haagen-Smit and in several of the many succeeding articles that relate aerosol inorganics to hydrocarbons and ozone in the production of photochemical smog.[2-4]

Second, nonmethane hydrocarbons are significant because of their physiological reaction with the human body.

Whereas such gases as methane and even some of its higher homologs have been statistically shown not to cause disease, many organic compounds such as tertiary amines, aldehydes, and oxides, together with the polycyclic aromatic hydrocarbons, have been shown to contribute to such diseases as emphysema and lung cancer.[5] Some of these compounds are listed in Table 1 together with their relationship to disease in man.

In truth, the estimation and evaluation of organic atmospheric emission is an extremely complicated undertaking. Detrimental effects are not caused by hydrocarbons directly, but by their derivatives that are produced by aerosol

Table 1 Relationship between Exposure to Contaminant and Corresponding Disease

Compound or Class of Compounds	Disease
Benzene	Anemia, leukopenia
Soots, tars, oils	Lung and skin cancer
Benzidine	Cancer of the urinary tract
Tertiary aromatic amines	Cancer of the bladder and urinary tract
2-Naphthylamine	Urinary irritation and cancer
Benz(*e*)acridines	Cancer in animals

reaction with other substances. Therefore, an appraisal of the effects of hydrocarbon pollution must be made by tracing the contaminant as measured through the "road map" of its aerosol reactions to its source. For this reason, it is necessary that we examine closely the sources of various classes of hydrocarbons to obtain a better understanding of hydrocarbon synergism and thus a more balanced appraisal of the atmosphere to be sampled.

CLASSES OF ORGANIC AIR CONTAMINANTS

Hydrocarbons

Strictly speaking, hydrocarbons are compounds whose molecules consist exclusively of carbon and hydrogen. Since the volatility of these compounds decreases with increasing molecular weight, one might expect that only the lighter hydrocarbons of C_1 to C_4 would be significant in air pollution. However, while the lower hydrocarbons exceed all others in quantity, the less common toxicants, although present in small quantities, are all the more dangerous as respiratory and general environmental contaminants. Such a liquid mixture as gasoline includes compounds of the butane and propane class and, at the same time, compounds that range from C_{12} to C_{20}.

In terms of their increasing detrimental effect on air quality, hydrocarbons can be arranged as follows:

1. Aliphatic.
2. Olefinic and aromatic.
3. Polycyclic aromatic.

Many subsidiary classifications are used to distinguish groups of hydrocarbons according to their reactivity, such as acyclic and alicyclic, but the three groups serve for the purposes of general sampling procedures at this point.

In fuel burning there is always the possibility of evaporation of some of the fuel constituents directly through the combustion process. In fact, the chromatogram of diesel exhaust closely resembles that of the fuel itself,[6] showing that, although combustion of fuel to CO_2 and water predominates, such physical effects as channeling of unburned fuel around the flame and quenching of the flame from the walls of the combustion channels are by no means rare occurrences.

Much information has been compiled by experiments that use gas chromatography to separate hydrocarbon components of the atmosphere for identification. Analysis of air collected in the early morning and later in the day near Riverside, California, by Stephens and Burleson[7] indicates that the more reactive hydrocarbons (underlined in Table 2) are greatly reduced after some hours of aging. The more reactive compounds, such as cis-2-butene and 1-butene, together with other olefins, showed marked decreases, indicating the devouring effect of aerosol reactions (1), (2), and (5). Substituted aliphatics, such as 3-methylpentane, exhibited only slight reduction in concentration, but inert aerosol compounds, such as methane and ethane, actually increased

Table 2 Hydrocarbon Concentrations in Morning and Afternoon Ambient Air Samples, Riverside, California, Fall 1968

Hydrocarbon	Concentration[7] (ppm)	
	7:30 A.M. PST September 24, 1968	4:10 P.M. PST October 24, 1968
Methane	2.3550	2.5300
Ethane	0.0636	0.0722
Ethylene	0.0636	0.0179
1-Butene	0.0026	0.0003
cis-2-Butene	0.0014	<0.0002
Butane	0.0276	0.0620
Isopentane	0.0392	0.0412
Cyclopentane	0.0032	0.0024
3-Methylpentane	0.0084	0.0061
Hexane	0.0090	0.0083
2-Methyl-2-butene	0.0026	<0.0002

slightly in concentration, as would be expected from a rather constant emission with increasing factor of time.

The detailed extent of these data suggests that analysis of atmospheres is perforce highly selective and requires elaborate equipment. If an exhaustive profile of air contaminants were required to effectively monitor ambient air, this would certainly be the case. For practical purposes, however, the analysis can be described as the product of two alternatives:

1. A rather comprehensive breakdown of ambient atmospheres can be accomplished by measuring a certain number of molecular components, using gas chromatography or a combination of gas chromatography and mass spectrography. Then, by means of the concentration kinetics of Benson,[8] which describe as a function of time the instantaneous level of contaminant as a function of the reaction in which it is consumed, one can calculate the profile of hydrocarbons at a given moment to a reasonable degree of accuracy.

2. Total reactive hydrocarbons can be determined by isolating methane and then subtracting the methane equivalent from the total hydrocarbons as found by flame ionization photometry, using a charcoal filter and relatively simple instrumentation.

Community air contamination from hydrocarbons is thus seen to follow a reproducible diurnal variation, as exemplified by Figures 1 through 3, which show the data collected by air sampling studies of the Los Angeles County Air Pollution Control laboratory.

Furthermore, on an annual basis there is a remarkable similarity between levels of nonmethane hydrocarbons, which appear to be independent of location of sampling site, and seasonal effects (Figures 4, 5, and 6). Consequently, any variance in nonmethane hydrocarbon corresponding to episodes of photochemical smog seems quite effectively monitored using the index of nonmethane hydrocarbons as defined in alternative 2 above.

Although it has proved difficult to develop an exact model of the hydrocarbon–smog relationship, an empirical approach has been adopted by the Environmental Protection Agency, Air Pollution Control Office (EPA-APCO), which is based on the measurement of the early morning levels of nonmethane hydrocarbon as an indication of potential for smog formation on a given day. This approach is valid since, by the definition implied in alternative 1, a statistical prediction of the degree of active hydrocarbon consumption can be made with acceptable accuracy using the kinetics of Benson; thus to prevent the oxidant from rising above 0.1 ppm for 1 hr, the early morning (6:00 to 9:00 A.M.) levels of nonmethane hydrocarbon must not exceed 0.3 ppm as methane.[9]

We have seen that measurement of nonmethane hydrocarbons is measurement of the photoreactive portion of this class of compounds. Since the analysis

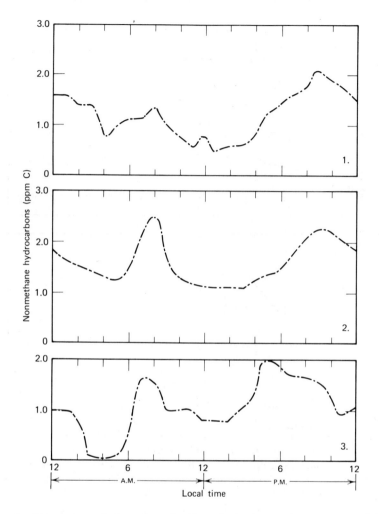

Figure 1. Nonmethane hydrocarbons by flame-ionization analyzer as annual hour averages for West Los Angeles. (From reference 10.)

Figure 2. Nonmethane hydrocarbons by flame-ionization analyzer as annual hour averages for Los Angeles. (From reference 10.)

Figure 3. Nonmethane hydrocarbons by flame-ionization analyzer as annual hour averages for Pasadena. (From reference 10.)

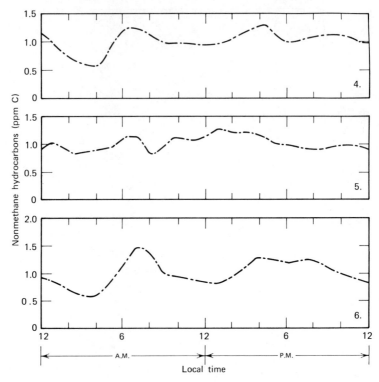

Figure 4. Nonmethane hydrocarbons by flame-ionization method as hour averages for Chicago (May to August and October 1968). (From reference 10.)
Figure 5. Nonmethane hydrocarbons by flame-ionization method as hour averages for St. Louis (May–July, September, and October 1968). (From reference 10.)
Figure 6. Nonmethane hydrocarbons by flame-ionization method as hour averages for Denver (January–March, May, September, and October 1968). (From reference 10.)

of air demands that we analyze it coincidently for a variety of contaminants, we must recognize as covariables the compounds that react with active hydrocarbons to yield a group of equally important and necessarily determinable nonmethane derivatives. A review of the substances into which hydrocarbons are converted through oxidation and substitutions is therefore in order.

Oxygenates and Hydrocarbon Derivatives

Aldehydes

Where mild oxidation of alcohols and epoxides is known to produce alcohols in the liquid phase, because of interfacial air oxidation in the aerosol phase, aldehydes arise as one of the major products in the photooxidation of hydrocarbons after their reaction with ozone, oxygen, or free radical intermediates such as hydroxyl or carbonium ions.

Once formed, the aldehyde RCHO group is less stable than other carbonyl nuclei, because the polarity of the $-C=O$ bond makes the carbon-bound hydrogen susceptible to nucleophilic attack. This is important because it is through this mechanism that other more photoreactive acyl and alkyl radicals are formed. Here

$$RCHO + OH \rightarrow RCO + HOH$$
$$\downarrow A$$
$$R + CO + A$$

where A is some photoreactive intermediate.

Consequently, during periods of high-flux density of sunlight, concentrations of aldehydes may reach 0.2 to 0.35 ppm in ambient air.[12,13]

Probably the most frequently encountered aldehyde is formaldehyde; its potential as an eye irritant is high, but not so high as that of acrolein (CH_2CHCHO) or peroxyacyl nitrates, which are also found as normal photochemical air contaminants. While formaldehyde usually makes up about one-half of the total weight of aldehydes in air, acrolein accounts for about 5%,[14,15] with the balance distributed between acetaldehyde and notably benzaldehyde which is an important constituent of automotive exhaust.[16]

Peroxyacyl Nitrates

Chemical aerosol breakdown of these aldehydes by means of mechanisms involving nitrogen oxide molecules and radicals produces the peroxyacyl group of air contaminants. An example of such a reaction

$$CH_3CO(OO) + NO \rightarrow RCO-OO-NO$$

shows the final stage of the aldehyde oxidation product in reaction with nitrogen dioxide to form the most common member of this group the peroxyacetylnitrate (PAN) itself.

Other derivatives often encountered in ambient air mixture are the acetyl and butyl nitrates from acetaldehyde and butyraldehyde or from reactions of nitrogen oxides with ethylene, propene, and butenes.[17]

Probably the most comprehensive early review of such photoreactions is that by Altshuller and Bufalini published in 1965,[18] which shows the stoichiometry of aldehydes, ketones, and alkyl derivatives but affirms the lack of data in reactions of intermediates with aryl aromatic compounds. Since then, some work using irradiated exhaust-containing atmospheres has shown the existence of peroxybenzylnitrate (PBN), which suggests that this compound is 1000 times greater a lacrimator than PAN.

In terms of the relative ability of various heavy hydrocarbons to yield peroxyacetylnitrate in irradiated atmospheres, the following are arranged in order of decreasing photochemical reactivity: *trans*-3-hexene, *cis*-3-hexene, and *o*-xylene. In general, although the lower hydrocarbons are more reactive in producing photoirritants, the heavier hydrocarbons produce compounds that are more irritating per mole.

SOURCES OF ORGANIC AIR CONTAMINANTS

Probably the largest single source of atmospheric hydrocarbons is the earth itself. The presence of natural hydrocarbons[19] has been known for some time and has been estimated to be 0.8 to 1.1 mg/m^3 (1.2 to 1.5 ppm).[20] Samples taken at various points on the globe show the estimated rate of emission of this contaminant from coal and peat fields, forest fires, natural coniferous respiration (Appalachian "Blue Haze" and other sources) to be in the neighborhood of 3×10^8 tons/year.[21] Assuming an instantaneous mass of 1×10^{13} lb to be present in the atmosphere at some moment, and assuming average climatic changes and radiation flux, Kozama in 1963 calculated an average atmospheric life of 20 years for such a mass, by using kinetics of atmospheric oxidation reactions. Others have since calculated shorter average lives for methane[22,23]; however, the implication persists that methane is relatively atmospherically unreactive.

Although natural sources contribute the most methane, a summary of hydrocarbon emissions from industrial and metropolitan areas in the United States indicates a substantial yield of hydrocarbons from normal mobile and stationary sources as a growth function of modern industrial technology. Of these emissions, it is estimated that transportation accounts for approximately 50% of the total weight of aerosol hydrocarbon, while industrial and residential fuel burning, together with industrial process evaporation, accounts for 14 to 16%. Approximately 26% of the total hydrocarbons comes from miscellaneous sources, such as coal, refuse, and solvent evaporation.[24]

Among miscellaneous sources of importance are the following:

Oil refineries—benzene, cresylic acid, phenol, anthracene, propane, furfural, thiols, thiophene.

Coking—naphthalene, anthracene, benzo(*a*)pyrene, cresylic acid, phenol.

Pharmaceutical industries—alcohols, ethers, esters, ketone, carbon disulfide.

Solvent dewaxing and degreasing—acetone, dichloroethylene, trichlorethylene, methylene chloride, perchlorethylene, methyl ethyl ketone, orthodichlorobenzene.

Dry cleaning industry—C_2 to C_5 chlorinated aliphatic hydrocarbons.

Textile industries—hexylene glycol, hydrocarbons, naphthalene, amines, acetates.

Food processing—furfurals, hydrocarbon oils, trimethylamine.

Chemical plastics industries—carbon disulfide, glycerol, methanol, isopropanol, isocyanates, amines, thiols, phthalic anhydride, naphthenes.

Paint preparation and application—esters, ketone, methyl and butyl Cellosolve, alcohols, hydrocarbons.

It should be noted that while most hydrocarbon solvents that enter the air through evaporation appear as gases or vapors, others, such as the naphthalenes, produce true aerosols. Some aerosol particulate matter such as PAN or PBN is formed in the atmosphere as are other peroxide, hydroperoxide, and aldehyde air contaminants. These are called secondary aerosols.

GENERAL METHODS OF ANALYSIS FOR ORGANIC AIR CONTAMINANTS

Several instrumental techniques can be used to measure total combustible hydrocarbon organics, but the analyst who is interested in analyzing for one or only a few specific compounds must to a large extent develop his own analytical approach. Several specialized reviews offer guidelines in general analytical techniques.[25,26]

Sample Collection

The method of sample collection must be designed to entrap particles whose average diameter, $<10\ \mu$, is of a size that permits their entry into the human lung. This can be accomplished by the standard high-volume fiberglass filter or a freeze-out trap (described in Chapter 5). A charcoal column is also a suitable collection medium, but chemisorption of the collected hydrocarbon may preclude the recovery of as much as 20% of the collected material.

Whereas a freeze-out trap collects fine particulate together with water vapor, a glass fiber filter allows collection of a dry sample. If further drying is necessary because of humidity of known process conditions, however, desiccator drying is preferred to oven drying, which could result in the loss of some of the volatile organic particulate.

Separation of Sample from the Collecting Medium

Once the sample has been collected, the desired portion of organic material is separated from the fiber by selective extraction, or a total organic particulate is extracted as a gross contaminant by mixed solvent extraction and reported as total organic hydrocarbon.

The benzene-soluble portion of organic particulate is frequently reported in documents from the Air Pollution Control Office of the Environmental Protection Agency, since this portion contains most of the "tars,"[27] epoxides,[28] and polycyclic aromatics[29] that have been experimentally related to the incidence of cancer. The measurement of this portion of organics has a long history in many large cities; annual data can thus be compared.

Selective solvent extraction using polar and/or nonpolar solvent systems can be accomplished as in Table 3. A more complete selection of solvent systems can be found in the text of Shriner and Fuson.[30]

Here, a generalization can be made. Almost all of the compounds of interest as normal air pollutants can be found in the fraction obtained by extraction with ether using dilute acid, and step 1 followed by dilute base, and step 2 to separate all inorganics and other interfering substances. This final fraction, step 3, according to Hoffman and Wynder,[31] makes up approximately 73.5% of the total extractable organics, leaving behind only the carboxylates and the amine portion of the air contaminant organics. As an example of a separation scheme, consider the approach developed by Hoffman and Wynder (Figure 7).

Step 1. Using the scheme shown in Figure 7, separate the organic ether insolubles by placing the following in a Soxhlet extractor:

1. Weighed (0.3 to 0.5 g) dry particulate sample.
2. Cut-up high-volume filter (use one-half of filter for single-sample extraction).

Table 3 Selective Solvent Systems for Extraction or Organic Contaminant from Dustfall or High-Volume Particulate

Solvent	
	Alcohols (ROH), some esters (RCOOR′), ketones (RCOR′), glycols (RCOHCH$_2$COHR′), and some glycol esters
Dil(3 N)HCl or 2 N H$_2$SO$_4$	Weak and strong carboxylic acids (RCOOH), amines (RNH$_2$)
2 N NaOH	Phenols (ArOH) and cresols [Ar(OH)$_3$]
NaOH(aq) + ether	Aliphatics, aromatics, ketones, esters, ethers
Benzene–methanol or benzene–isopropanol	Most organics, polycyclic aromatics, chlorinated hydrocarbons, and tars (high-molecular-weight aromatic and polycyclic bituminous particulate)
Benzene	Most aromatic compounds, polycyclic aromatics, and tars
Cyclohexane	Most aliphatic and polycyclic aromatic compounds

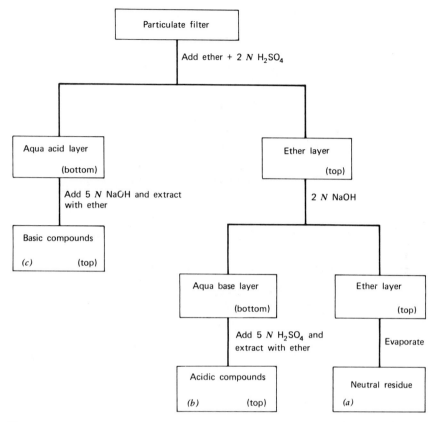

Figure 7. Scheme for separation of organics from high volume or dustfall sampling medium. (a) Here the "neutral residue" contains saturated, unsaturated, aromatic, oxygenated, halogenated, polycyclic aromatic and miscellaneous nonpolar compounds. (b) The acidic compounds represent nonvolatile carboxylic acids, RCOOH. (c) The basic compounds are amines together with N-heterocyclic aromatic hydrocarbons.

3. Two-inch portions[31] of five to seven high-volume filters as a composite sample.

4. The weighed residue from freeze-out trap.

The extracting is done with ether for a period of 4 hr at a low temperature, taking care to prevent loss of solvent. (*Note.* Concentration of the ether extract through overheating could produce explosive peroxides.) The ether extract may then be transferred to a 125-ml separatory funnel (after filtering, if necessary).

If a Soxhlet extractor is not available, the weighed samples may be placed in a 50-ml beaker, combined with small portions of ether (use adequate ventilation), and stirred until only insoluble material remains. The sample may then be filtered through a "medium" sintered glass funnel into a 125-ml separatory funnel.

Step 2. The total ether extract is then shaken with a 15 to 25 ml portion of 2 N H_2SO_4. If only the neutral organic portion is desired, the aqueous portion of the separatory funnel can be made basic with sodium hydroxide pellets, adding four to six pellets before testing the pH of the solution by means of pH or litmus paper.

Disgarding the aqueous layer, which may contain a small amount of precipitate, wash the ether layer with three successive 25-ml portions of distilled water, add a small amount (5 to 10 g) of anhydrous sodium or calcium sulfate, and allow to stand for 24 hr to remove any absorbed water.

The ether-soluble neutrals may then be evaporated to dryness at room temperature, using some vacuum and warming slightly, and then desiccator dried and weighed in a tared 25-ml beaker or evaporating dish.

Step 3. This neutral fraction is then separated into aliphatic, aromatic, and oxygenated constituents by column chromatography.

1. Fill a ½-in. column with chromatographic grade silica gel, condition the column with 50 ml of isooctane, and add (all at once) the total neutral fraction, which has been prepared by mixing to a thick slurry with silica gel and isooctane. Elute with a volume of isooctane that is about three times the capacity of the column. Collect the aliphatics in a small beaker and allow the solution to evaporate, transferring to a 25-ml tared beaker for final weighing.

2. Following the separation of aliphatics, elute the column with approximately three times its volume of chromatographic grade benzene.*

3. After the separation of aliphatic and aromatic constituents the column is eluted with approximately three times its volume of a mixture of chromatographic grade 1:1 methanol and chloroform. Evaporate at room temperature and weigh the oxygenated and substituted hydrocarbons, which contain ethers, ketones, and esters.

The procedures just described are standard methods of organic separation, as found in a number of works, which may be used for further reference.[30,32]

Using the procedure above as a general format, other solvents can be substituted in the primary separation steps. For example, according to Liberti et al.,[33] both polar and nonpolar organic constituents in a particulate sample can be separated by using 4:1 methanol and cyclohexane. Here the residue from

---

* It is especially important to use chromatographic UV grade benzene, which should be redistilled just before use. If spectroscopy will be employed to distinguish aromatic compounds, spectroscopic grade benzene should be used for this elution.

evaporation of the cyclohexane layer can be extracted with nitromethane to yield a concentrated solution of the polycyclic hydrocarbons in the nitromethane layer. For convenience, the extraction can be combined into one step when the water–methanol layer is heavier than cyclohexane, but lighter than nitromethane; a three-layer system may be shaken and separated into polar compounds, aromatics, and polycyclics, respectively.

Analysis of Separated Organic Fractions

Flame Tests for Chlorinated Hydrocarbons

Chlorinated hydrocarbons, which are found in (step 3), fractions (1) and (2), may be detected in parts per million by gathering a few grains of the weighed fraction on the tip of a hydrocarbon-moistened copper wire and inserting in the flame to produce a green color, if Cl is present (Beilstein test).

Determination of Aromatic Compounds by Refractometry

If cyclohexane is used to extract the total organics, the total aromatic content can be estimated simply and with rather good accuracy by using relatively small volumes of reagents. If, after extraction, a volume of 0.5 ml of organics remains, this amount can be used to wet the prisms of an Abbe refractometer and the dispersion index; that is, the index of refraction for the F (hydrogen) line minus the index of refraction for the C (hydrogen) line, divided by the density can be used as a rapid direct estimate of the total aromatic content.[34]

Determination of Aromatic and Olefinic Compounds by UV Absorption

Since many aromatic compounds absorb in the ultraviolet region, UV absorption spectrophotometry can be used to identify and quantitatively assay specific aromatics such as substituted benzenes and some polycyclic aromatic compounds. Here the isooctane fraction can serve as a suitable sample in which to measure directly the UV absorption. Pentane is also quite transparent in the UV region.

The peak UV absorption of several aromatics is given in Table 4 to demonstrate the degree of resolution required to separate mixtures of common solvents. However, the same compounds, if present singly, can be determined quite simply by using the UV absorption maximum to compare standards in isooctane.

A double-beam spectrophotometer allows a minimum solvent interference where, as the proximity of the absorption maximum in Table 4 indicates, solvent purity of the isooctane is desirable and a spectrophotometer of good

Table 4 UV Wavelength Absorption Maxima of Several Common Organic Solvent Pollutants

Compound	Wavelength Units Absorption Maximum (MW)
Toluene	262
Benzene	262
o-Xylene	263
p-Xylene	274
m-Xylene	275

resolution is a necessity. Standards are prepared and usual Beer-Lambert law calculations are used to arrive at concentration values.

Determination of Functional Group Organics by Infrared Absorption

Whereas UV absorption is used to determine most of the aromatic compounds, infrared absorption is employed to qualitatively identify also functional group (oxygenated, chlorinated) and aliphatic compounds, once a good extraction separation had been achieved (e.g., if the mixture of organic contaminant is not unusually complex); it is also used in cases where sample separation and retention time information have been accomplished by gas chromatography.

Probably the most effective use of infrared analysis in air pollution chemistry is in solvent analysis, where *pure* standards are available and identification can be made by comparing known spectra to samples collected by freeze-out or impinger techniques. Paint solvents, paint baking effluent, volatiles from polymer synthesis or plastics fabrication processes, and some specific petroleum refinery vapors can be introduced into 0.025, 0.1, or 1.0 mm liquid cells, or into 250-ml gas cells in the case of highly volatile materials. When nonvolatile materials such as tars, heavy drawing oils, or even water-soluble materials from industrial processes are being investigated, a mull or "smear" may be prepared with Nujol. Nujol is then mixed 10:1 with the purified particulate and is wiped onto the surface of demountable infrared cells for comparison with a Nujol-blank path. To investigate water-soluble crystalline particulate, a technique has been developed that uses potassium bromide, in which the solid is mixed in weighed proportion with KBr and high-pressure-formed into very thin infrared translucent pellets, which can be mounted in the sample path of the infrared spectrophotometer.

OTHER MEANS FOR IDENTIFICATION AND QUANTITATIVE ANALYSIS OF SPECIFIC ORGANIC AIR CONTAMINANTS

Identification and Analysis of Aromatic Compounds

Analysis by π Complex Reactions

As the π-basicity (i.e., electron-donating ability) of an aromatic compound increases, the color of the compound formed between the aromatic and a reagent π-acid compound such as tetracyanoethylene,[35,36] piperonal,[37] or 2,4,7-trinitrofluoranone[38] varies from red, through green and blue, to violet. This chromophoric behavior can be used as a basis for either qualitative or quantitative analysis.

Introduction

Depending on the complexity of a particular mixture, the method of π-complexation can be applied with varying degrees of success. Like infrared spectrometry, these methods are most successful when the sample is relatively pure. The reliability of these tests should be confirmed by performing the test on a sample of the particular contaminant from a known industrial or process source. Once identification is confirmed in the presence of whatever process interferences are present, quantitation can be performed by preparing Beer's law plots in the presence of the known interference at concentrations equivalent to the process ratios. High sample purity is required, but once that pure standard has been obtained, steps to achieve sample purification can be checked against control mixtures, which include the standard, to ensure adequate recovery of the desired constituent after the purification process is complete.

While most unsaturated and even aliphatic functional group compounds react with one or more reagent π-acids, the Lewis basicity of the aromatic π-system often produces such strong absorption of energy that the resulting transmitted color lies outside the visible region of the human eye. Therefore, a given π-acid may not produce a suitable colorimetric qualitation test, although UV quantitation may be achieved if the analyst has the time and instrumentation to pursue fullfield UV spectrophotometry scans of various "colorless" π-acid adducts of this desired compound.

In summary, π-acids (electron pair acceptors) such as tetracyanoethylene, piperonal, and 2,4,7-trinitrofluoranone can be used to colorimetrically identify and spectrophotometrically determine concentrations of polycyclic aromatic, aromatic, aromatic amines and phenols, and some aliphatic ketones by either direct or quenching techniques.

The reader may wish to try analysis by quenching when direct analysis of an isolable contaminant appears difficult. Compounds such as phenanthrene, anthracene, and trinitrobenzene can, because of their electron system, absorb

70   Sources and Measurement of Organic Air Contaminants

the fluorescence emission from otherwise visibly fluorescent compounds, many of which are air contaminants of the pyrene class. A thorough treatment of this phenomenon is beyond the scope of this book, but a number of references are available on the subject.[39] Also, some familiarity with this approach can be achieved by adding a few grains of tetracyanoethylene to a dilute solution of pyrene in methylene chloride to observe the marked decrease in fluorescence.

Since these phenomena take place in the chemically inert polyringed systems where spectroscopy of some form will almost always be involved in the determination, an added variable such as emission quenching can prove to be very handy.

The Piperonal Test[40]

Collect a sample on a high-volume filter extract for 8 hr with benzene using a Soxhlet extractor, evaporate it to dryness at room temperature with a stream of nitrogen, and take up the resulting residue with 10 ml of chloroform.

Add one drop of the chloroform solution to one drop of a reagent prepared by dissolving 5% piperonal and 6.8% phosphorous pentachloride in chloroform (stable for up to 3 hr). Add 10 drops of trifluoroacetic acid to a small test tube or spot plate containing this mixture, mix thoroughly, and allow to stand for 5 min to observe color change. A red, purple-blue, green, or violet color confirms the presence of aromatics. Note that with small samples or with a fixed sample of particulate this test can be performed simply under the microscope, as described in Chapter 2. A partial list of aromatic colors is given in Table 5.

To quantitate the reaction above, add 1.0 ml of the chloroform solution of the desired aromatic compound to a 10-ml volumetric flask. Add 1.0 ml of the "reagent," dilute to volume with trichoroacetin acid, mix, and allow to stand for 15 min before reading the color intensity. Color is stable for about 30 min. The exact absorption maximum for the compound should be determined by an absorbance scan from 400 to 800 mμ in a double-beam spectrophotometer versus a reagent blank. Ability to quantitate varies from compound to compound; consequently, expected interferences should be introduced into a control sample to permit assessment of their energy absorbance at the λ maximum of the desired constituent.

According to Sawicki,[41] a K value can then be determined:

$$K = \frac{E}{m} \times \frac{W_t}{W_a}$$

where E = absorbance (in a 1-cm cell) of the solution prepared in trichloroacetic acid from a gram sample weight W_t of total benzene-soluble fraction of which an aliquot weight (grams) sample W_a was taken for analysis. This sample then represents m cubic meters of air (obtained by multiplying the cubic meters per minute flow rate of sampling by the time in minutes).

Table 5 Color and Absorption Maximum Wavelength for Complexes of Piperonal with Organics Including Common Air Contaminants[40]

Compound	Color	λ Maximum (mμ)
Pyrene	Green	680
Benzo(a)pyrene	Yellow-brown	755
Anthranene	Yellow-brown	792
Fluoranthene	Blue	625
Xylene	Red	522
9-Methylanthracene	Red	737
Phenyl selenide	Blue-green	612
Methyl phenyl sulfide	Purple	575
9-Ethylcarbazole	Blue	582
Indole	Red	502
Biphenyl	Violet	560
Fluorene	Purple	580
Benzo(e)pyrene	Green	705

Reprinted with permission from E. R. Sawicki, R. Miller, T. W. Stanley and T. Hauser, "Detection of Polynuclear Hydrocarbons and Phenols with Benzol and Piperonal Chlorides," *Anal. Chem.*, **30**, 1130 (1958). Copyright by the American Chemical Society.

The definition of K then follows: "The specific absorption coefficient of organic matter from 1000 m³ of air per 100 ml of treated solution."[42]

The K values thus obtained are proportional to the amounts of various aromatic compounds present in the community air. A plot of K values versus the wavelengths scanned produces the variance shown in Figure 8.

Knowing the absorption maxima of various compounds from Table 5 and having perhaps added to this list some compounds of his own, the analyst is in a position to "diagnose" the organic composition of the air in his own or any other community by using the general rule that *the complexity of the absorbing aromatics' ring system increases with increasing wavelength of absorption maximum.*

Thus in Figure 8, where the air sample from Birmingham shows large amounts of the fluorene and naphthalene systems, both cities were found to yield only small amounts of the simple benzene aromatics and to show a decline in the multiringed perylene polycyclics. Subsequent analysis of these samples by spectrofluorometry disclosed the respective concentration of benzo(a)pyrene in the ratio of 8.2:1 for Birmingham versus Los Angeles. The K values from

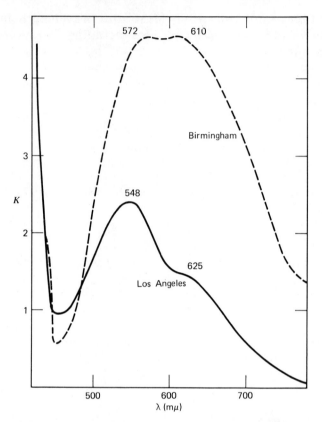

Figure 8. The K value versus the wavelength in millimicrons. Piperonal test of the benzene-soluble fraction of Los Angeles (———) and Birmingham (-----) air particulates.[41] (From reference 41.)

the graph at 755 mμ are observed to follow this proportion with the same order of magnitude.

The sensitivity of this test is shown by the fact that dust collected from a laboratory 2 weeks after synthesis of the serious carcinogen N-2-fluorenylacetamide on piperonal test yielded a blue solution having an absorption maximum at 593 mμ, where the listed λ maximum of pure N-2-fluorenylacetamide is 595 mμ.

Thus the piperonal test, together with the K versus λ scan plot of an air particulate sample, provides a rapid means of estimating the distribution of aromatic compounds in community air.

The 3-Nitro-4-dimethylaminobenzaldehyde Reaction[43]

Phosphorus pentachloride acts as an intermediate in the electrophilic substitution of large multiple-ringed organics to produce colors in the near-infrared and visible region. While compounds such as benzene, toluene, fluorene, naphthalene, and biphenyl that yield complexes with TCNE and piperonal are not conjugated enough to permit reaction with 3-NDB, other compounds such as the benzo(*ghi*)perylenes are found to react quantitatively.

The procedure here is similar to the piperonal test, although 0.5 ml of phosphorus oxychloride is added to 1.0 ml each of reagent and chloroform extract before diluting to 10 ml with trifluoroacetic acid. The reagent is prepared by dissolving 50 mg of the pentachloride in 20 ml of *o*-dichlorobenzene. This mixture is then heated to boiling to clear and is cooled before diluting to 25 ml with *o*-dichlorobenzene.

The compounds that yield positive results with this procedure are listed in the work of Sawicki and Barry.[44.]

Other Methods for Aromatics

Among the less general tests for aromatics are the Isatin tests,[45] in which colored compounds are formed with several multiringed aromatics, Deniges' test[46] for benzene, and the iodine colorimetry of Benesi and Hildebrand,[47] which is somewhat less reliable and less applicable than the above-mentioned procedures when dealing with the small amounts of sample that are frequently used in air analysis.

Although the piperonal test has not been widely accepted,[48] partly because of the growth of microfluorometry with development of lower cost filter instruments, it is still used when survey approximation is desired.

Analysis of Polynuclear Aromatic Hydrocarbons (PAH)

Introduction

The most widely used method for the separation and quantitation of this type of aromatic aerosols includes benzene or cyclohexane extraction of the high-volume filter, or some portion thereof, followed by thin-layer chromatographic (TLC) development and subsequent spectrophotofluorometric determination of the desired TLC separated compounds.

Each step in this sequence of procedures must be carefully standardized so that a percentage recovery factor can be assigned not only to the preliminary extraction, but also to the TLC separation and specific fluorescent activity of each compound in the fluorometer. Such percentage recoveries, once obtained, are quite reproducible and must act as weight coefficients to be employed in determining actual unknown amounts of an organic by the same procedure.

The compounds given in Figure 9 (together with fluorescence emission spectra) commonly make up this polycyclic aromatic fraction and have been identified in urban air samples.

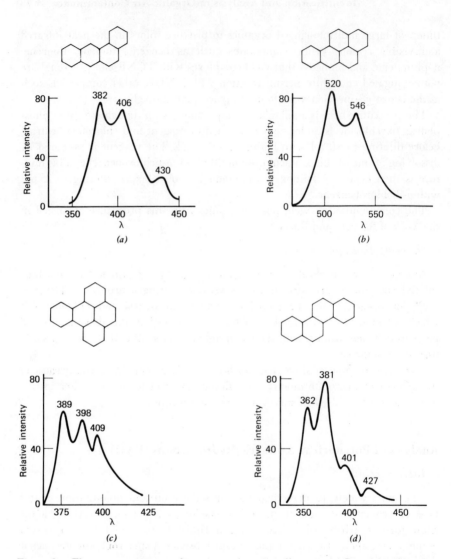

Figure 9. Fluorescence spectra of some common air pollutants. (a) Benz(a)anthracene: Concentration, 5 × 10^{-5} M in pentane; activation λ, 284 mμ. (b) Benz(a)pyrene: Concentration, 5 × 10^{-7} M in sulfuric acid; activation λ, 520 mμ. (c) Benz(e)pyrene: Concentration, 5 × 10^{-8} M in pentane; activation λ, 330 mμ. (d) Chrysene. (e) Perylene: Concentration, 5 × 10^{-8} M in pentane; activation λ, 430 mμ. (f) Anthanthrene: Concentration, 5 × 10^{-8} M in pentane; activation λ, 420 mμ. (g) Naphtho[2,3](a)pyrene: Concentration, 5 × 10^{-7} M in pentane; activation λ, 450 mμ. (h) Dibenzo(e,l)pyrene. (i) Dibenzo(a,i)pyrene. (j) Indeno(1,2,3-cd)pyrene. (k) Dibenz(a,h)acridene. Other polycyclics: (l) 7H-Benz(de)anthrancene-7-one; (m) Phenalen-1-one.

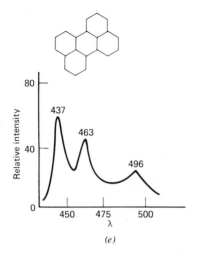
(e)

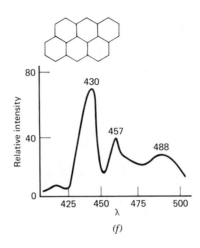
(f)

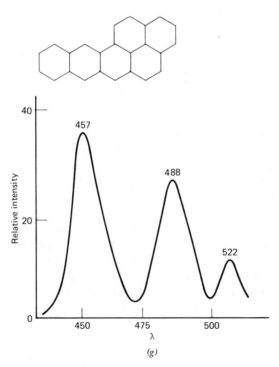
(g)

Figure 9. (*Continued.*)

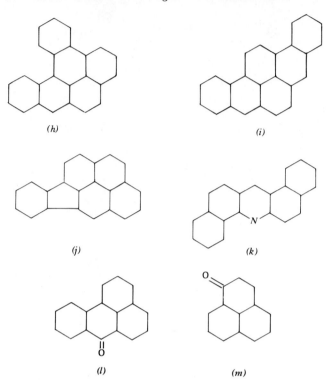

Figure 9. (Continued.)

Of these, anthanthrene and phenylen-1-one have been identified as aerosol products of coffee roasting. Others arise from incinerators that use refuse, plastics, and sewage as a charge. Some noncombustion sources are asphalt batch and coking operations—up to 5 $\mu g/m^3$ of benzo(a)pyrene has been found in the vicinity of coking ovens.[49]

The temperature required to burn organic matter easily breaks the C—C and C—H bonds to produce free radicals. These radicals then cyclize to the lowest energy ring system allowable with respect to the number of active sites, the degree of unsaturation, and the number of existing carbon atoms.

In the fluorescence analysis of these compounds, consistently high fluorescence intensities all but guarantee detection in microgram to nanogram quantities. Once the sample has been separated by TLC, a sulfuric acid solution of the B(a)P spot activation at 520 mμ permits fluorescence detectable to 0.01 μg.[50]

Procedure for fluorescent analysis of the polycyclic aromatic hydrocarbon fraction of the benzene or cyclohexane soluble portion of suspended particulate includes sample preparation as follows.

Sample Preparation

Ordinarily 16 in.² of exposed fiberglass filter paper is allotted for analysis of organic compounds. Although detectable limits for most compounds permit the use of such a small sample, often for convenience several strips will be extracted in the same Soxhlet extraction thimble to produce a composite residue representing 1 week out of perhaps a month sampling at the same site. Such a composite is recommended when only average ambient levels are desired and daily data variations need not be shown (Figure 10).

To a large extent, TLC has replaced paper and column chromatography as a means for separating individual polycyclics before fluorescence activation, although the TLC separated B(a)P fraction often contains traces of chrysenes and benzo(k)fluoranthrene, which must be separated by spectrofluor selectivity.

While benzene is the most popular solvent for this preliminary extraction,[52] Monkman[53] and others have found the cyclohexane extract to be much freer of asphaltic tars and other amorphous material that otherwise quenches fluorescence and causes "tailing" in the development of TLC plates.

As a compromise, preliminary benezene extraction may be performed, which upon evaporation to dryness may be taken up in cyclohexane before thin-layer spotting is begun.

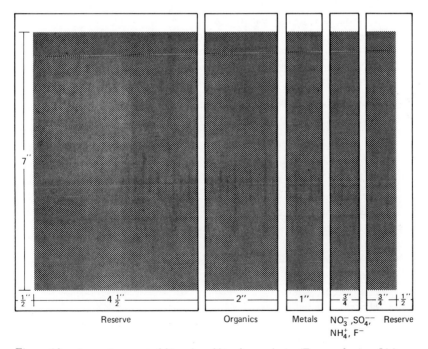

Figure 10. Apportionment of fiberglass filter for analysis. (From reference 51.)

Procedure for Extraction of PAH's

One or more fiberglass filter strips, a weighed dry sample obtained by electrostatic precipitator, or a residue from freeze-out collection is placed in the thimble of a Soxhlet extractor and allowed to cycle continuously for 8 hr. Ordinarily 95 to 99% of the benzene-soluble residue is separated in this step.

The extract is then evaporated to approximately 10 ml over a water bath at $90 \pm 5°C$. A quantitative transfer is made to a tared 5-ml beaker that has been placed in a 50-ml heating mantle. To prevent oxidation of the polynuclear aromatic hydrocarbons (PNAH), the evaporation is continued under a stream of nitrogen. A thermometer is also placed (bulb down) in the heating mantle to ensure a temperature of $90 \pm 5°C$. This step in the procedure requires constant attention, since the temperature tends to rise immediately after evaporation of the organic solvent. Also, the evaporation must be accomplished under a hood, and care must be taken because of the flammability of these organic solvents.

Procedure for the TLC Separation of PAH's

Using the residue from the benzene–cyclohexane extraction, Sawicki et al.[54] evaluated a number of methods for the determination of B(a)P, including several TLC procedures for further separation of the PNAH. Two methods have been frequently mentioned, one of which uses a thin-layer plate of alumina–cellulose acetate (2:1) with ethanol–toluene–water (17:4:4) as the developer while the other employs an alumina plate with pentane–ether (19:1) as solvent. Both procedures, however, involve scraping of the developed spots from the thin-layer plate, which is a tedious experience. Alternatively, the separation may be performed by Gelman Instant Thin-Layer Chromatography Type SG (ITLC). In the commercial preparation of these sheets, silica gel absorbents are impregnated into the glass microfibers. While there are many advantages in the use of ITLC, the separation of the tar fraction is especially significant: (1) solvent migration is very rapid, that is, 8 to 10 min rather than the 40 to 60 min with regular TLC; and (2) the spots can be cut out with scissors eliminating the laborious scraping procedure. One disadvantage in using ITLC is a lack of clearly defined spot separation. If one is very careful in cutting out the developed spots, reproducible results can be obtained.[55] Let us continue to assume that the ITLC procedure is to be followed.

The weighed organic residue of extraction (at least 60 μg of particulate from 6 m^3 of urban atmosphere) is dissolved in 0.1 ml of benzene, and a 0.1 to 1 mg aliquot is spotted 1.5 cm from the bottom of the ITLC sheet that has been activated at 105°C for 1 hr. A reference line should be drawn 1.5 cm from the bottom of the ITLC sheet before activation to ensure that all samples will travel an equal distance during development. A reference spot containing 0.5 to 5 μg of each analytically desired contaminant should be run on each ITLC sheet. Three unknown samples and one reference may be spotted individually on each 20 x 20 cm sheet.

When applying the organic residue aliquot to the ITLC sheet, each spot should be kept as small as possible to ensure its own wholeness. The simplest method is by means of a capillary tube. However, a micropipette is preferred, since the volume spotted is accurately known; 10 to 15 λ pipettes allow the spot to remain small and the solvent to dry between applications. Of course, an appropriate aliquot for ITLC may be transferred to a small beaker and a capillary tube used to spot the sample on an activated sheet.

After the ITLC sheet has been spotted and the solvent evaporated, the sheet is placed in TLC equilibrium chamber. The ITLC sheet allows one to use a sandwich chamber and to expect full one-dimensional development in 10 min or less.

In analyzing for benzo(a)pyrene, the cyclohexane–benzene extract is further evaporated on ITLC, using n-pentane for development. The solvent front is allowed to travel to about 0.5 cm from the top of the sheet, at which time the sheet is removed and permitted to dry in an area shielded from UV light.

After drying, a UV lamp is allowed to irradiate the ITLC sheet so that the desired spots are clearly defined for marking before they are removed with sharp scissors. Whether long or shortwave UV illumination is used depends on which constituent is involved. Thus B(a)P exhibits greater fluorescence on ITLC with the short wavelength lamp.

When the desired spots have been defined, they are cut out from the sheet with scissors that have been rinsed in 6 N HNO_3. Each spot is then cut into very fine squares and placed separately in test tubes that have been rinsed in 6 N HNO_3 and dried in an oven. These spots are then extracted with diethyl ether. Generally, the extraction is complete after four rinsings; however, it is suggested that the trimmings be checked for fluorescence by using an UV lamp. (Be careful to note possible fluorescence of impurities in the ether.)

The ether extracts are then combined and evaporated over a tepid water bath *under a hood.* If the extract cannot be evaporated to dryness over the water bath, the test tube is placed in a 50-ml heating mantle surrounded by glass wool. A Variac setting 5 to 30 V is preferred for very gentle heat, and a nitrogen atmosphere is run into the test tube to prevent oxidation.

Spectrofluorometric Determination of PAH's

The eluted B(a)P spot is then transferred to a 10 or 25-ml Erlenmeyer flask and swirled slowly in 2.0 ml of concentrated H_2SO_4 (sp. gr. 1.84) until dissolved. When most or all of the air bubbles have been released, the sample solution is transferred to a spectrophotofluorometer for measurement of concentration. It is convenient to add 2.0 ml of H_2SO_4, since 1 ml is usually insufficient to ensure an adequate volume in the 1-cm quartz cell.

When the number of air bubbles has decreased considerably (after no more than 4 to 5 min of swirling), the $B(a)P$ H_2SO_4 complex is transferred to a 1-cm quartz cell that has been thoroughly cleaned, rinsed in 6 N HNO_3, and oven-dried. The cell is held carefully at the top and all four sides are wiped free of fingerprints, lint, and so on. (Kimwipe or some other lintfree material is recommended.) It is important to remember that an abundance of air bubbles and/or foreign matter on the inside or outside of the cell wall will cause heterogeneous scattering attenuation of light and lead to analytical error. After inserting the cell into the cell holder, $B(a)P$ is determined at an activation (excitation) wavelength of 525 mm and a fluorescence (emission) wavelength of 545 mm, that is, F 525/545. While Sawicki advocated F 470/545 as well, other investigators of the spectroscopy of $B(a)P$ also recommend F 525/545.[55] Purity of the compound can be checked by running an activation spectrum. Figure 11 is the activation spectrum of 5×10^{-6} M $B(a)P$ at a fluorescence λ of 545 mμ in sulfuric acid.

The Aminco-Bowman spectrophotofluorometer is to date perhaps the most widely used instrument for the measurement of PNAH and has found continued acceptance in air pollution analysis. Several models have been developed, each necessitating various adaptations of slit arrangements, smaller

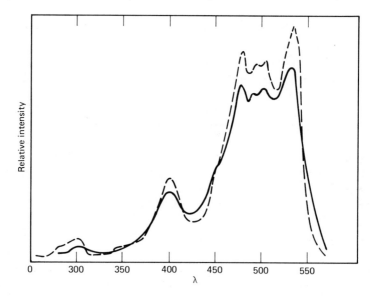

Figure 11. Activation spectra in concentrated H_2SO_4 at fluorescence λ of 545 mμ. Dashed line: 5×10^{-6} M BaP. Solid line: BaP spot from ITLC.

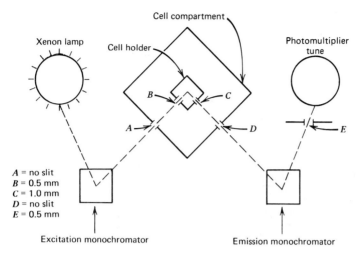

Figure 12. Slit arrangement for Aminco-Bowman spectrophotofluorometer.

slit widths giving greater resolution and larger slit widths allowing more sensitivity. The slit arrangement for the determination of B(*a*)P in older models of this instrument is reported by Sawicki,[54] while in the more recent model,[56] the disposition illustrated in Figure 12 can be used.

Other PNAH compounds are also determined spectrophotofluorometrically,[57] although their role as suspected environmental carcinogens is as yet unknown.

ANALYSIS OF OXYGENATED AND MISCELLANEOUS SUBSTITUTED HYDROCARBONS

The neutral oxygenated group of hydrocarbons shown in the separation scheme in Figure 7 are found in air in amounts varying from a few tenths to several parts per million, depending on the proximity of a source—the spraying of automotive parts, or garbage incineration, for example.

Infrared scan of the neutral oxygenated fraction of air-collectible particulate or vapor most often yields the carbonyl absorption wavelength (stretching)

$$\overset{|}{\underset{|}{C}} = O$$

which lies between 5.4 and 6.2 μ.

Ketones

In an adaptation of the well-known method for the identification and quantitative determination of ketones,[58] a hypodermic needle is filled with a 10% solution of 2,4-dinitrophenylhydrazine in 4:1 ethanol–water. This is then used to spot a fiberglass mat on which ketone has been collected in the form of microscopic particulate or adsorbed vapor.

This reagent is prepared by weighing 10 g of the 2,4-dinitrophenylhydrazine and dissolving it in 15 ml of concentrated sulfuric acid. This solubilized solution is then very slowly stirred into a mixture of 70 ml of 95% ethanol and 20 ml of water.

A glass fiber filter or a piece of filter paper is then used as the test medium. The ketone should be partially extracted by adding a drop of 95% ethanol to the mat or paper; a drop of the test reagent will then produce a color, which may appear immediately or may require up to 10 min to develop. The color indicates which predominant ketones are present. The dinitrophenylhydrazones of ketones, which are unconjugated with a double bond or contain a functional group, are yellow, while aromatic aliphatic of functional group conjugation shifts the absorption band to produce an orange or red color.

This test is sensitive to 0.1 to 0.2 ml of ketone. Examples of conjugated versus unconjugated positive colors are the following: acetone, yellow; methyl isobutyl ketone, yellow; mesityl oxide, red.

Since some aldehydes yield a positive test with this reagent and since certain alcohols, especially allyl alcohol and its derivatives, may be oxidized to aldehydes on standing in contact with the reagent (or merely the ambient air), care should be taken to allow for this possible interference in the evaluation of results. In any case, a negative test rules out the presence of carbonyls in general, while a positive test indicates the need for further confirmation by gas chromatography, using known ketones of the suspected class as control species.

Another test employs methyl Cellosolve[59] (ethylene glycol monomethyl ether) with anhydrous aluminum chloride, in which substitution of the ketone (which is not directly connected to a ring) produces a blue color whose wavelength maximum lies in the 595 to 605 mμ region of the spectrum. Sensitivity is in the order of 1 to 5 μg.

Procedure for Ketones That Are Not Directly Ring-Substituted

Using a small piece of glass fiber mat or filter paper on which sample has been collected, under a microscope add a microdrop of methyl Cellosolve and five microdrops of 0.5%, 2-hydroxyl-1-naphthaldehyde in methanol. Allow HCl gas (from an open bottle of concentrated HCl) to pass through the spot for a few seconds. Five microdrops of o-dichlorobenzene is then added, although the spot may have already turned a green color if the paper contains a high concentration of ketone. The spot may then be oven heated at 50°C for a few minutes, or until the odor of methanol is no longer noticeable. If examination with the microscope confirms the appearance of a green color that disappears on heating but recurs on cooling, a positive test is recorded.[60] It is recommended that this test be rehearsed using a ketone, such as 2-butanone, 2-heptanone, or 1-phenylpropanone, which yields positive color formation.

Aldehydes

Probably the most prominent carbonyl found in ambient air, caused by use of the internal combustion engine as a source of power, is formaldehyde. It is therefore necessary to have on hand a specific test for formaldehyde in addition to a method for determining of total aldehydes.

Procedure for Formaldehyde Using Chromotropic Acid[61]

The sample can be collected directly in the color-forming reagent impingers, using midget impingers or bubblers as described in Chapter 3. Two bubblers are arranged to allow an airflow of 1 ft^3/min through a solution prepared by dissolving 1.0 g of chromotropic acid in 1000 ml of concentrated sulfuric acid.

After a sampling time of 30 min the impinger solution is transferred to a boiling water bath for 10 min to ensure complete reaction with development of color. Cool and measure the absorbance at 580 mμ versus a reagent blank that has been similarly treated in the boiling water bath.

A Beer-Lambert plot of known concentrations can serve to determine the formaldehyde concentration in parts per million, using the known air sample volume as calculated from the airflow velocity and sample collection time.

The color of the reaction product remains stable for some time. However, although it is effective in determining formaldehyde in most industrial atmospheres, this method suffers interference from glucose as well as several less commonly encountered organic condensation products of functional group hydrocarbons.

Procedure for Total Aldehydes Using MBTH[62]

Approximately 35 ml of 0.05% aqueous 3-methyl-2-benzothiazolone hydrazone hydrochloride (MBTH) is placed in each of two impingers in series, and 1 ft^3/min of sample air is drawn through the solution for 30 min. If ambient air is of low relative humidity, a small amount of water may be required to adjust the impinger volume to 35 ml after sampling is completed and before leaving the solution at room temperature for 1 hr.

After 1 hr, when the reaction is completed, pipette 10 ml of each test solution into two 15-ml test tubes and add 2 ml of an aqueous filtered solution that contains 1.6% sulfamic acid and 1% ferric chloride. The contents of the two are combined and allowed to stand for 10 to 15 min before taking a reading at 628 mµ versus a reagent blank made up of 10 ml MBTH reagent to which 2 ml of the sulfamic acid–ferric chloride solution has been added.

The parts per million of the total aldehyde are then calculated by the formula

$$\text{ppm} = \frac{21.4a}{v}$$

where a is the absorbance and v is the sampling volume in liters.

To determine acrolein in such atmospheres as photochemical smog, where it acts as an extreme lacrimator, Cohen and Altschuller have established a method using 4-hexylresorcinal which involves only the preparation of a "mixed reagent," into which ambient air is bubbled.

Procedure for Acrolein Using 4-Hexylresorcinol[63]

Prepare stock solutions as follows:

1. Using a 250-ml beaker in a heated water bath, dissolve 200 g of trichloracetic acid in 10 ml of distilled water.
2. Dissolve 3.0 g of mercuric chloride in 100 ml of 95% ethanol.
3. Dissolve 5 g of 4-hexylresorcinol in 5.5 ml of 95% ethanol.

A "mixed reagent" of the 4-hexylresorcinol is now prepared by combining 25 ml of ethanol with 250 ml of solution 1, 1 ml of solution 2, and 0.2 ml of solution 3.

Ten milliliters of this "mixed reagent" is then placed in each of two impingers or bubblers. Sample air is drawn through the bubblers, which are placed in series, at the rate of 1 ft 3 min for 30 min. The collecting solution in each bubbler is then transferred to test tubes and heated in a water bath for 15 min at 60°C.

Absorbances are read colorimetrically in a double-beam spectrophotometer at 605 mµ versus a reagent blank. Any absorbance in the sample from the second bubbler should be added to the first for a combined absorbance A.

A standard 5 to 10 liter bag can be used to prepare standards from which a Beer-Lambert curve can be drawn, or the equation

$$\text{ppm} = \frac{8.7a}{v}$$

where a is the total absorbance and v is the sample air volume in liters, can be used to calculate the parts per million of acrolein.

These tests are best standardized by preparing a standard 5 to 10 liter bag using a known concentration of desired carbonyl and vaporizing the liquid into the bag by way of a known flow of air, as described in Chapter 4 on calibration procedures.

Where benzaldehyde and other aromatic or polycyclic aldehydes are found to be the emission products of nonleaded, higher aromatic blended, so-called pollution free gasolines, a test for this class of carbonyls is desirable to ensure knowledge of ambient air quality during a time of changeover in automotive technology.

Knowledge of the level of this aromatic fraction of total aldehydes is especially important in urban areas, where automotive commuting is observed to elevate air contaminant levels between hours of 7:00 and 9:00 A.M. and 4:00 and 7:00 P.M. each day.

Procedure for the Estimation of Aromatic and Polycyclic Aldehydes

The aromatic aldehyde is dissolved from a glass fiber filter, or separated as a vapor from a freeze-out trap or some other collecting medium, using chloroform.

In a 15-ml test tube, 1 ml of a 5% fluoranthene in chloroform solution is mixed with an equal amount of the chloroform–aldehyde containing extract. With constant mixing, 0.8 ml of trifluoracetic anhydride is added, and the mixture is allowed to stand for 5 min. The sample is diluted with 10 ml trifluoroacetic acid and optical density is read at the λ absorption maximum in a double-beam spectrophotometer versus a reagent blank. Color remains stable for several hours.

Since the test depends on the available π electron system of the aromatic ring, *o*-, *m*-, or *p*-nitrobenzaldehydes and the aromatics that are substituted with high electron-withdrawing groups fail to react.

Table 6 lists some aromatics and their test results as reported by Sawicki.[64]

In the past the most frequently used test for lower-molecular-weight al-

Table 6 Colors of Aromatic Aldehyde Reaction Products after Treatment with Fluoranthene in Trichloroacetic Acid[64]

Aldehyde	λ Maximum (mµ)	Color
Benzaldehyde	590 (cloudy)	Purple-blue
Vanillin	612	Blue
1-Naphthaldehyde	660	Blue
Mesitaldehyde	607	Blue
2-Naphthaldehyde	615	Blue
o-Phthaldehyde	477	Orange
1-Pyranaldehyde	752	Green

dehydes and for the methyl ketones has been the bisulfite test for carbonyls.[65] Other methods include the reaction of aldehydes with phenylhydrazine (Schiffs reagent),[66,67] a test that is often used in industrial hygiene practices to detect "in-plant" aldehydes caused by small-vehicle exhaust emission.

Esters

Since many lacquer, enamel, and spray polish operations use commercially available coatings that employ esters such as ethyl acetate and isobutyl acetate as solvents, it is important to use a test that will permit their quantitative determination in the community as a whole. Usually it is the air concentration in the vicinity of the source that is tested for suspected solvents, rather than general ambient air analysis being done.

Esters are identified by treating either the freeze-out sample of vapor or the chloroform extract from a glass fiber high-volume sampler, first with an alkaline solution of hydroxylamine hydrochloride and, after 10 min, with an aqueous solution of 10% perchlorate. A violet color indicates the presence of esters.

Alternatively, the alkaline solution of hydroxylamine hydrochloride can be used as the collecting reagent in a bubbler that is operated at 1 ft^3/min for 15 to 30 min. Once identified in the mixture (where the detection limit varies from 0.04 to 0.10 μl of ester), the ester may be traced in gas chromatograph effluent by using a 25 to 50 ml hypodermic syringe to collect suspected peak effluents. The effluent gas is then bubbled into alkaline hydroxylamine with the subsequent formation of the hydroxamic color. Quantitation may then proceed, using the peak area of the ester.

Other Organic Compounds

The syringe technique just discussed can be used to identify other gas chromatograph peaks, with quantitation being carried out on the chromatograph; known carbonyls are used to establish concentration-intensity calibration curves.

Aliphatic Alcohols

Aliphatic alcohols can be collected in the chromatograph and bubbled into a ceric ammonium nitrate solution. Most lower weight alcohols such as ethanol, methanol, or isopropanol produce an amber color, while the higher aliphatic alcohols appear orange to red.[68]

Chlorinated Hydrocarbons

The syringe method may also be adapted to the Beilstein test for chlorinated hydrocarbons. The effluent from the gas chromatograph is introduced directly into the Bunsen burner flame. A green color indicates the presence of chlorinated hydrocarbons to a detection limit of 0.13 to 0.20 μl.[69]

Olefins

Olefins in concentration less than 5 ppm can be determined by reaction with p-dimethylaminobenzaldehyde using the method of Altschuller and Sleva.[70]

Procedure for Determination of Olefins

Into each of two impingers or bubblers pipette 20 ml of concentrated sulfuric acid and 0.25 ml of a reagent prepared by dissolving 5 g of p-dimethylaminobenzaldehyde in 42.9 ml of glacial acetic acid.

Ambient air is drawn through the bubblers for 15 min, after which the contents of the bubblers are transferred to a single 50-ml test tube (or are distributed between two 25-ml test tubes), placed in a boiling-water bath, and heated for 20 min. The samples are then cooled, and absorbance is read at 500 mμ versus a reagent blank.

In final calculations, a standard parts per million sample can be prepared in a Tedlar bag and dilutions used to establish a standard Beer-Lambert concentration curve; or the olefin content can be given by the formula

$$\mu g/\text{liter} = \frac{143a}{v}$$

where a is the absorbance and v is the volume of the air sampled, in liters.

Still other organic compounds such as anthracene,[71] pyrene,[72] o-, and p-quinones,[73] and fluorenes[74] can be determined by methods noted in the literature.

THE GAS CHROMATOGRAPH AS A MEANS FOR DETERMINING ORGANIC AIR CONTAMINANTS

Needless to say, the most versatile instrument for the determination of the widest possible spectrum of organic air contaminants is the gas chromatograph, especially if it has been fitted with a flame ionization as well as a thermal detector.

This combination permits detection of trace quantities of hydrocarbons (FID) without sacrificing the capability of detecting higher concentrations of functional group hydrocarbons such as the aldehydes and flame-inert compounds such as nitrogen, carbon monoxide, and helium (thermal detector).

For air pollution work, a suitable column, prepared by depositing 5% 1,2,3-tris(2-cyanoethoxy)propane on 60 to 80 mesh acid-washed firebrick, has been proposed by Clemons, Leach, and Altshuller.[75]

While very low-molecular-weight hydrocarbons such as methane and ethane are separated rather well at room temperature, heavy hydrocarbons such as naphthas and heavy aliphatic solvent materials frequently require a 150 to 200°C column temperature.

A heated sampling loop is equally useful, since actual gas or vapor samples can be introduced directly into the column without recourse to condensing methods to concentrate the sample for liquid injection.

Procedure for Gas Chromatographic Analysis

Analysis of hydrocarbon mixtures can be undertaken using gas chromatography to separate and quantitatively determine mixtures of hydrocarbons in the low part per million range. This approach is effective in that quantitative rather than qualitative information is obtained, and the most successful pollution analysis is the one in which the air effluents are *known* hydrocarbon products, which may be introduced into the chromatograph for use as standards.

In using the gas chromatograph, the flow of nitrogen or helium carrier gas is set to 100 ml/min, and the column temperature is set between 50 and 150°C, as determined by the expected molecular weight of hydrocarbons in the sample.

Preparation of Calibration Standards

Liquid hydrocarbon standards can be used to prepare a standard gas mixture as long as the exact density of the liquid is known.

A calibration mixture can be prepared from several common air contaminants, and the volume concentration versus peak area relationships can be used to quantitate data for any of the components subsequently encountered in ambient air samples.

Using a 10-μl pipette, transfer 3.0 μl of benzene, 4.0 μl of toluene, and 5.0 μl of *o*-xylene to a Tedlar bag by sweeping with clean air as described in Chapter 5, or dissolve 30, 40, and 50 μl, respectively, in 10-ml carbon tetrachloride, or carbon disulfide and transfer a 1-ml aliquot of this mixture to the bag.

The actual gaseous volume of these liquids can be calculated from the following relationship: μl of vapor volume = μl of liquid hydrocarbons $[(d \times MV)/MW] \times 10^3$ where d = density of the liquid, MV = 24.4 μl/μmole (molar volume of any gas at 25°C), and MW = molecular weight. Using an airflow of 10 liters/min and a stopwatch to determine 10.0-min elapsed time, fill to a volume of 100 liters a Tedlar bag with "clean" air into which the hydrocarbons are vaporized.

At 10 liters/min flush the hydrocarbon standards into a Tedlar bag, allowing a 10.0-min flow to produce a final volume of 100 liters. Concentrations of individual hydrocarbons can then be determined by means of the equation given above:

$$\frac{\mu \text{l of vapor volume (HC)}}{\text{total volume in liters}} = \text{ppm (by volume)}$$

Fill the gas loop of the chromatograph and inject the sample. Since CCl_4 or CS_2 exit almost immediately, retention times of the standard hydrocarbons may be determined for the equation above, where the order of elution is benzene, toluene, and o-xylene. Once peaks are identified and labeled in this order and retention times recorded along with column and flow parameters, peak areas are used to produce a quantitative standard.

Gaseous samples are introduced into the chromatograph by filling the gas loop and injecting its entire contents into the column at time zero. Liquids are injected using a microliter syringe; a 1 to 5 μl volume syringe is quite useful for most samples. Retention times are noted for the various peaks that occur on the readout of the strip chart recorder.

Relative concentrations can be estimated from peak heights and suspected knowns can then be introduced to establish peak area concentration correspondence in parts per million. The traced shape of a peak can be weighed and the area calculated by comparison to the weight of a known peak. Alternatively, a rough area can be calculated by the average area method, using

$$a = \tfrac{1}{2} (w_b \times l_h f_a)$$

where w_b is the width of the peak at the base, l_h is the peak height, and f_a is the actual attenuation factor used to produce a peak within the span of the chart.

A library of known retention times at specific carrier flow and column temperature should be kept as part of any laboratory record. A suggested list of such compounds should include benzene, toluene, o-, m-, and p-xylene, cumene, aliphatics from C_6 to C_{12}, chlorinated aliphatics and o-dichlorobenzene, vm and p solvents, Stoddard solvent, and p-cymene.

Silicone or 1540 Carbowax columns are usually adequate to separate general hydrocarbon mixtures; however, a commercially available column has been prepared by Clemons, Leach, and Altshuller specifically for the chromatographic analysis of hydrocarbons. The stationary phase here is 1,2,3-tris(2-cyanoethoxy)propane.

REFERENCES

1. R. J. Cvetanovic, "Addition of Atoms to Olefins in the Gas Phase," in *Advances in Photochemistry*, W. A. Noyes, Jr., G. S. Hammond, and J. N. Pitts, eds., Vol. 1, Interscience, New York, 1963, pp. 115–182.
2. A. J. Haagen-Smit and M. M. Fox, "Ozone Formation in Photochemical Oxidation of Organic Substances," *Ind. Eng. Chem.*, **48**, 1484–1487 (September 1956).
3. E. R. Stephens et al., "Photochemical Reaction Products in Air Pollution," *Int. J. Air Water Poll.*, **4**, 79–100 (June 1961).
4. J. J. Bufalini and A. P. Altshuller, "Synergistic Effects in the Photooxidation of Mixed Hydrocarbons," *Environ. Sci. Technol.*, **1**, 133–138 (February 1967).
5. L. Greenburg et al., "Benzene Poisoning in the Rotogravure Printing Industry in New York City," *J. Ind. Hyg. Toxicol.*, **21**, (8), 395 (1939).
6. D. F. Merrion, "Effect of Design Revisions on Two Stroke Cycle Diesel Engine Exhaust," *SAE Trans.*, 1968.
7. E. R. Stephens, and F. R. Burleson, "Distribution of Light Hydrocarbons in Ambient Air," presented, 62nd Annual Meeting of Air Pollution Control Association, June 22–26, 1969.
8. S. W. Benson, *Thermochemical Kinetics*, Wiley, New York, 1968.
9. "Air Quality Criteria for Hydrocarbons," *op. cit.*, Section 5.
10. "Air Quality Criteria for Hydrocarbons," *op. cit.*, Section 3, p. 5.
11. "Air Quality Criteria for Hydrocarbons," *op. cit.*, Section 3, p. 4.
12. A. P. Altshuller et al., "Photochemical Reactivities of Paraffinic Hydrocarbon-Nitrogen Oxide Mixtures Upon Addition of Propylene or Toluene," *J. Air Poll. Control Assoc.*, **19**, 792 (1969).
13. W. E. Scott and L. R. Reckner, "Atmospheric Reaction Studies in the Los Angeles Basin," Progress Report, November 15, 1968 to January 15, 1969, Scott Research Laboratory, APRAC Project, CAPA, 7-68 (1969).
14. N. A. Renyetti and R. J. Bryan, "Atmospheric Sampling for Aldehydes and Eye Irritation in Los Angeles Smog—1969," *J. Air Poll. Control Assoc.*, **11**, 421–424, 427 (1961).
15. A. P. Altshuller and S. P. McPherson, "Spectrophotometric Analysis of Aldehydes in the Los Angeles Atmosphere," *J. Air Poll. Control Assoc.*, **13**, 109–111 (1963).
16. D. F. Merrion, *op. cit.*, p. 68.

17. E. A. Schuck, G. J. Doyle, and J. Endow, "Photochemistry of Polluted Atmospheres," Standard Research Institute, July 1960.
18. A. P. Altshuller and J. J. Bufalini, "Photochemical Aspects of Air Pollution: A Review," *Photochem. Photobiol.*, **4**, 97–146 (1965).
19. M. V. Migeotte, "Spectroscopic Evidence of Methane in the Earth's Atmosphere," *Phys. Rev.*, **73**, (5), 519–520 (1948).
20. C. E. Junge, *Air Chemistry and Radioactivity*, International Geophysical Series, J. Van Meighem, ed., Vol. 4, Academic Press, New York, 1963, p. 382.
21. T. Koyama, "Gaseous Metabolism in Lake Sediments and Paddy Soils and Production of Atmospheric Methane and Hydrogen," *J. Geophys. Res.*, **63**, (13), 3971–3983 (1963).
22. D. H. Ehhalt, "Methane in the Atmosphere," *J. Air Poll. Control Assoc.*, **17**, 518–519 (1967).
23. E. Robinson and R. C. Robbins, "Sources Abundance and Fate of Gaseous Atmospheric Pollutants," Stanford Research Institute, February 1968.
24. National Air Pollution Control Administration Reference Book of Nationwide Emissions, U.S. Department of DHEW, PHS, CPEHS, NAPCA, Durham, N.C.
25. A. P. Altshuller, "Analysis of Organic Pollutants," in *Air Pollution*, A. C. Stern, ed., Vol. 2, Academic Press, New York, 1968, pp. 115–145.
26. M. Shepard et al., "Isolation, Identification and Estimation of Gaseous Pollutants in Air," *Anal. Chem.*, **23**, 1431–1440 (1951).
27. R. Schoental, *Polycyclic Hydrocarbons*, E. Clar, ed., Vol. 1, Academic Press, New York, 1964, p. 487.
28. H. L. Falk, P. Kotin, and A. Mehler, "Polycyclic Hydrocarbons as Carcinogens for Man," *Arch. Environ., Health*, **8**, 721–730 (May 1964).
29. P. Kotin and H. L. Falk, "Atmospheric Factors in the Pathogenesis of Lung Cancer," *Adv. Cancer Res.*, **7**, 475 (1963).
30. R. L. Shriner, R. C. Fuson, and D. Y. Curtin, *The Systematic Identification of Organic Compounds*, Wiley, New York, 1965.
31. D. Hoffman and E. L. Wynder, "Organic Particulate Pollutants," in *Air Pollution*, A. C. Stern, ed., Vol. 2, p. 191, Academic Press, New York, 1968.
32. N. D. Cheronis and J. B. Entrikin, *Semimicro Qualitative Organic Analysis*, Crowell, New York, 1963.
33. A. Liberti et al., "Gas Chromatographic Determination Polynuclear Hydrocarbons in Dust," *J. Chromatogr.*, **15**, 141 (1964).
34. B. E. Saltzman and R. L. Larkin, "The Industrial Environment, Its Evaluation and Control," C. H. Powell and A. D. Hosey, eds., PHS Publication, B-18-7, Washington, D.C., 1965.
35. P. V. Peurifoy and M. Nager, "Sensitive Spot Test for Nitrogen Compounds in Petroleum Fractions," *Anal. Chem.*, **32**, 1135 (August 1960).

36. G. H. Schenk, P. O. Warner, and W. Bazzell, "Determination of Tertiary Aromatic Amines Using Tetracyanoethylene," *Anal. Chem.*, **38**, 907 (1966).
37. E. Sawicki, R. Miller, T. Stanely, and T. R. Hauser, "Detection of Polynuclear Hydrocarbons and Phenols with Benzol and Piperonal Chlorides," *Anal. Chem.*, **30**, 1130–1133 (1958).
38. G. H. Schenk et al., "Study of the Pi Complexes of 2,4,7-trinitrofluorenone with Phenols, Aromatic Hydrocarbons, and Aromatic Amines," *Anal. Chem.*, **37**, 372 (1965).
39. G. H. Schenk and N. Radke, "A Study of the Effect of Tetracyanoethylene, Trinitrofluorenone, and Other Pi Acceptors on the Fluorescence of Aromatic Carbons," *Anal. Chem.*, **37**, 910 (1965).
40. E. R. Sawicki, R. Miller, T. W. Stanley, and T. Hauser, "Detection of Polynuclear Hydrocarbons and Phenols with Benzol and Piperonal Chlorides," *Anal. Chem.*, **30**, 1130 (1958).
41. E. Sawicki, T. W. Stanley, and T. Hauser, "Detection of Heterosubstituted Aromatic Derivatives and Determination of Aromatics in the Air," *Chem. Anal.*, **47**, 69 (1958).
42. E. Sawicki, "Analysis of Aromatic Organic Compounds," unpublished analysis of atmospheric organics, Training Course Manual in Air Pollution, PHS, HEW, NAPCA, 1969.
43. *Ibid.*
44. E. Sawicki and R. Barry, "New Color Tests for Larger Polynuclear Aromatic Compounds," *Talanta*, **2**, 128–134 (1959).
45. E. Sawicki, T. W. Stanley, T. R. Hauser, and R. Barry, "Isatin Tests for Aromatic Hydrocarbons and Phenols," *Anal. Chem.*, **31**, 1664–1667 (1959).
46. B. R. Hubbard and L. Silverman, "Rapid Method for Determination of Hydrocarbons in the Air," *J. Ind. Hyg. Occup. Med.*, **2**, 49–55 (1950).
47. H. A. Benesi and J. H. Hildebrand, "A Spectrophotometric Investigation of the Interaction of Iodine with Aromatic Hydrocarbons," *J. Am. Chem. Soc.*, **71**, 2703 (1949).
48. D. Hoffman and E. L. Wynder, "Organic Particulate Pollutants," in *Air Pollution*, A. C. Stern, ed., Vol. 2, p. 209, Academic Press, New York, p. 209.
49. J. O. Jackson, P. O. Warner, and T. Mooney, Jr., "Profiles of Benzo(a)pyrene and Coal Tar Pitch Volatiles at and in the Immediate Vicinity of a Coke Oven," *Am. Ind. Hyg. Assoc. J.*, **35**, 276 (1974).
50. *Ibid.*
51. R. J. Thompson, G. B. Morgan, and L. J. Purdue, "Analysis of Selected Elements in Atmospheric Particulate Matter by Atomic Absorption," *At. Absorpt. New*, **9**, 53 (1970).
52. E. C. Tabor, T. R. Hauser, and J. P. Lodge, "Characteristics of the Organic Particulate Matter in the Atmosphere of Certain American Cities," *Arch. Ind. Health*, **17**, 58 (1958).

References

53. J. L. Monkman, G. E. Moore, and M. Katz, "Analysis of Polycyclic Hydrocarbons in Particulate Pollutants," *Am. Ind. Hyg. Assoc. J.*, **23**, 487 (1962).
54. E. Sawicki et al., "Comparison of Methods for Determination of Benzo(a)Pyrene in Particulates from Urban and Other Atmospheres," *Atmos. Environ.*, **1**, 131–145 (1967).
55. Wayne County Air Pollution Control Laboratory, unpublished data, Detroit, Mich., 1971.
56. Aminco-Bowman Spectrofluorometer Instruction Manual, American Instrument Company, 1965.
57. S. Undenfriend, *Fluorescence Assay in Biology and Medicine,* Academic Press, New York, 1962.
58. R. L. Shriner, R. C. Fuson, and D. Y. Curtin, *The Systematic Identification of Organic Compounds,* Wiley, New York, 1965, p. 126.
59. E. Sawicki and T. W. Stanley, "A Simple, Sensitive Test for Aliphatic Ketones," *Anal. Chem.*, **31**, 122–124 (1959).
60. E. Sawicki, "Thermochromic Detection of Acetonyl Grouping in Organic Compounds," *Chem. Anal.*, **48**, 4 (1959).
61. P. W. West and B. Sen, "Spectrophotometric Determination of Traces of Formaldehyde," *Z. Anal. Chem.*, **153**, 177 (1956).
62. E. Sawicki et al., "The 3-Methyl-2-benzothiazolone hydrozone Test," *Anal. Chem.*, **33**, 92 (1961).
63. I. R. Cohen and A. P. Altshuller, "A New Spectrophotometric Method for the Determination of Acrolein in Combustion Gases and in the Atmosphere," *Anal. Chem.*, **33**, 726 (1961).
64. E. Sawicki, T. W. Stanley, and T. R. Hauser, "Detection of Aromatic Aldehydes," *Chem. Anal.*, **47**, 31 (1958).
65. "Laboratory Methods: Aldehydes," Los Angeles County Air Pollution Control District, California Method, 5-46, 1958.
66. G. R. Lyles, F. B. Dowling, and V. J. Blanchard, "Quantitative Determination of Formaldehyde in the Parts per Hundred Million Concentration," *J. Air Poll. Control Assoc.*, **15**, 106 (1965).
67. R. L. Shriner, R. C. Fuson, and D. Y. Curtin, *The Systematic Identification of Organic Compounds,* Wiley, New York, 1965, p. 147.
68. B. E. Saltzman and R. L. Larkin, "The Industrial Environment, Its Evaluation and Control," C. H. Powell and A. D. Hosey, eds., PHS Publication 614, Washington, D.C., 1965, p. B-18-9.
69. B. E. Saltzman, and R. L. Larkin, *op. cit.*, p. B-18-10.
70. A. P. Altshuller, L. J. Lage, and S. F. Sleva, "Determination of Olefins in Combustion Gases and in the Atmosphere," *Am. Ind. Hyg. Assoc. J.*, **23**, 289 (1962).
71. T. R. Hauser, "Thermochromic Detection of Anthracene Derivatives," *Chem. Anal.*, **48**, 86 (1959).

72. E. Sawicki and T. W. Stanley, "A Simple Specific Spot Test for Pyrene and Its Derivatives," *Chem. Anal.,* **49,** 34, (1960).
73. E. Sawicki, T. W. Stanley, and T. R. Hauser, "Spectra Detection of Terminal Ring Quinones," *Anal. Chem. Acta,* **21,** 392 (1959).
74. E. Sawicki and W. Elbert, "Thermochromic Detection of Polynuclear Compounds Containing the Fluorenic Methylene Group," *Chem. Anal.,* **48,** 68 (1959).
75. C. A. Clemons, P. W. Leach, and A. P. Altshuller, "1,2,3-Tris(2-cyanoethoxyl)propane as a Stationary Phase in the Gas Chromatographic Analysis of Aromatic Hydrocarbons," *Anal. Chem.,* **35,** 1546 (1963).

3

SOURCES AND MEASUREMENT OF INORGANIC AIR CONTAMINANTS

INTRODUCTION

Air, which is a mixture of gases, has the remarkable property of being consistent in composition up to an altitude of 7 mi from the surface of the earth. Within this region, the composition of the air envelope is 78.08% nitrogen, 20.95% oxygen, 0.034% carbon dioxide, and 0.93% argon, together with quite smaller amounts of hydrogen and the so-called rare gases–neon, krypton, xenon, and helium.

Of the many properties of air, its life-sustaining capacity, attributable to gaseous oxygen, is of greatest interest to the biologist. Any interference with this most necessary function because of the presence of other gases or particulates toxic to life is necessarily of interest to the ecologist and, ultimately, the air pollution chemist. These include carbon monoxide, carbon dioxide, sulfur compounds, nitrogen oxides, oxidants, peroxyacetyl nitrate, and metals and metal ions.

CARBON MONOXIDE (CO)

Man's experience with CO probably began with the discovery of fire. Generally, the materials used to build a fire for the purpose of producing heat are wood, coal, or other bituminous organic material, and more recently, natural or artificial gas. When the use of coal for domestic fuel became common in the Western Hemisphere, sometime after the fifteenth century, problems associated with man's exposure to CO began to enter the annals of history. Usual incidences of CO poisoning have resulted from exposure to the products of incomplete combustion of fuel, coal damp, or swamp gas, and, more recently, from careless venting of exhaust gases produced by internal combustion engines. The toxic effects of CO are well known, with acute toxicity being experienced at levels well in excess of 100 to 200 ppm. Below these levels, the long-term effects of CO inhalation are less documented.

Health Effects

Carbon monoxide reacts with blood hemoglobin to produce carboxyhemoglobin, thereby reducing the effectiveness of this oxygen (O_2) carrier in the bloodstream to form oxyhemoglobin. Consequently, cells that depend on oxygen for life maintenance are oxygen-starved. Because O_2 and CO are very much alike in molecular structure, they complete for the same sites in the hemoglobin molecule. Carbon monoxide poisoning does not take place all at once. In obedience to the laws governing attainment of equilibrium, such poisoning comes about after several hours of inhalation at moderate levels (200 to 1000 ppm). The distribution of CO in the bloodstream is a direct function of the partial pressure of CO in the atmosphere, and the equilibrium that proceeds according to

$$CO + \text{hemoglobin} \rightarrow \text{carboxyhemoglobin}$$

is shifted to the right until the partial pressure of CO in the bloodstream is equal to that of CO in ambient air. The time period during which symptoms such as flushing and rapid breathing occur may be thought of as a "coming to equilibrium" with the ambient concentration of carbon monoxide. During this exposure, a blood test may be employed to determine the actual *in vivo* effect of a subject's exposure.

Measurement of CO in Blood

Since the actual blood color is directly affected by the presence of oxidized or reduced iron in the hemoglobin, it is a rather simple matter to dilute blood samples with ammonium hydroxide. Then, according to the method of Amenta, read the absorbances at 575, 560, and 498 mμ.[1] These absorbances are then used to prepare an absorbance ratio, using known solutions of oxyhemoglobin and carboxyhemoglobin to yield percent carboxyhemoglobin colorimetrically.

Other methods for CO in the blood are aimed at the CO itself and involve a separation step designed to destroy the blood matrix, thus freeing any hemoglobin-bound CO for subsequent chemical analysis. Such methods, although quite accurate, are often at a disadvantage because of the size of the blood sample required for any manometric or volumetric measurement. Some success has been reported in the use of ordinary Draeger-type CO detector tubes to measure blood-derived CO[2], while the classical Van Slyke method, although tedious, produces excellent analytical accuracy.

In another reported method, a solution containing excess palladium chloride is treated with blood-liberated CO, which reduces the palladium chloride to free palladium.[3] Subsequent colorimetry is performed on the remaining palladium ion to yield an indirect measure of blood CO.

Of the most recent instrumental methods, nondispersive infrared[3] provides a quite specific method for CO, provided that all water vapor is removed during the blood outgassing. In addition, several gas chromatographic methods can be applied where liberated CO can be either converted to methane[4] or determined directly[5] after liberation from blood carboxyhemoglobin.

Blood Levels of CO as a Means of Estimating Ambient CO Concentrations

An interesting sidelight to the discussion of in-blood carboxyhemoglobin determinations is the indirect relationship employed by Peterson and Stewart, which yields equilibrium carboxyhemoglobin concentrations as a function of ambient CO levels when the duration of exposure to CO is known.[6]

$$\log \% \text{ carboxyhemoglobin} = 0.85753 \log CO + 0.62995 \log t - 2.29519$$

Of interest here is that the converse of this equation yields the log of ambient CO, so that the above carboxyhemoglobin test may be used as an indirect method of measuring the ambient exposure at which the human subject came to equilibrium.

Sources of Carbon Monoxide

Natural Sources

Carbon monoxide is found naturally in gases produced by volcanoes, coal deposits, and the decaying organic material commonly found in swamps or bogs. Other natural sources are the stratospheric decomposition of CO_2,[1] as well as atmospheric dissociation of CO_2 intermediates in the photochemical formation of smog.[8]

Although the reaction of $CO + O_2 \rightarrow CO_2$ is chemically possible, its effectiveness as a CO scavenger is poor. According to Bates and Witherspoon, physical considerations make this reaction as written with no intermediate seem highly unlikely. Nevertheless, this scavenging reaction does occur at temperatures above 500°F and at room temperatures in the presence of the catalyst called Hopcalite, which is composed of mixed manganese and copper oxide. This reaction forms the basis for a quantitative determination of CO using a bed of Hopcalite for conversion of CO to CO_2.[9]

Biological Sources

Leaves of some plants,[10] together with colonies of hydrozoan jellyfish and various other oceanographic biological species,[11] have been shown to be CO producers along with man himself. Man's CO output varies from 10 to 90 ppm in breath, depending on factors such as degree of physical exercise, smoking habits, and general metabolic makeup.[12]

Combustion Sources

As one might expect, elevated levels of CO outside the home are generally found near expressways and at points where traffic density is highest.[13] This situation is apparent from the data shown in Figure 1. Other references to observations of CO as a function of automobile traffic can be found in the works of Brief et al.[14]

In general, it is not surprising to find express-apron and at-the-curb downtown street concentrations of CO at parts per million levels of three to four times that of residential areas. In the latter, background levels of CO are generally in the neighborhood of 0.1 ppm.[15]

An estimated 58%[16] of the national CO emissions are produced by motor vehicles as exhaust emittant. Other sources of CO emission include the combustion of coal, oil, and natural gas as well as combustion-related sources (e.g., coking of coal or asphaltic oil and the incineration of solid waste).

Miscellaneous Sources

Other miscellaneous noncombustion related sources of CO include the manu-

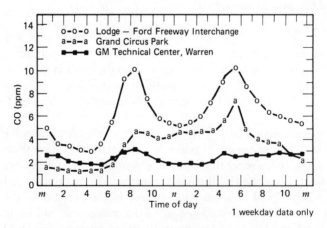

Figure 1. Time patterns of hourly average carbon monoxide variations in Detroit area sampling sites. [Reprinted with permission from J. M. Colucci, and C. R. Begeman "Carbon Monoxide in Detroit, New York and Los Angeles Air," *Environ. Sci. Tech.*, **3**, 41 (1969) Copyright by the American Chemical Society].

Carbon Monoxide (CO)   99

facture of chemicals such as organic acids, ketones, formaldehyde, and methanol, in addition to solvent and chemical production associated with catalytic petroleum cracking. Carbon monoxide is also produced in foundry operations when molten iron is poured into castings that have been lined with a layer of carbon-black as conditioning material.

Removal of CO from the Atmosphere

In the past several years, the scientific community has witnessed the publication of numerous articles speculating on possible mechanisms for removal of CO from the atmosphere. If it were not for one or more such "sinks," which apparently act to control the turnover of CO in the ambient air, our estimated U.S. emissions of over 5 million tons per year[17] would be expected to raise the observed world background from 0.1 ppm by as much as 0.03 ppm/year.[18]

The scientific bases for these "CO sinks" include known reactions of CO with H_2 or H_2O in the presence of microorganisms or bacteria, such as methanobacterium formicum, to form CH_4,[19] and, in some cases, lactic acid.[20] In addition, by somewhat similar mechanisms, vegetation may act as an important link in removing CO from the environment by binding CO molecules into porphyrin-type compounds.

Methods of Measurement

Absorption Method

Classically, CO is measured either by absorption or combustion. The most common absorbant is cuprous chloride, which forms the complex $Cu_2Cl_2 \cdot 2CO \cdot 4H_2O$ in either acidic or alkaline solution. The method using an alkaline solution has met with greater success. Absorption is accomplished using the Orsat or Bureau of Mines apparatus in which CO is bubbled through a chamber containing the cuprous chloride solution.[21] Recent improvements of this method include the use of Cosorbant, a commercially available absorbant. Cosorbant employs a mixture of β-napthol and cuprous chloride to slow absorption of atmosphere oxygen, which would otherwise interfere.

The absorption method usually involves measurement of the change in volume of reagent resulting from the reaction of CO to form a complex. Consequently, only moderate-to-high concentrations (100 ppm) of CO can be determined with confidence.

The advantage of absorptimetry lies in the practical ease with which air or stack gas containing CO can be drawn through the apparatus and CO estimated quickly from the change in volume of the absorbing solution. Practical experience, however, limits the accuracy of this method to concentrations above 10 ppm.

Iodine Pentoxide Method

The iodine pentoxide method is perhaps the chemical approach best suited for determining small amounts of CO. The reaction is

$$I_2O_5 + 5CO = I_2 + 5CO_2$$

Accuracy of the results depends on removal of such interferences as unsaturated hydrocarbons, water vapor, and oxides of nitrogen.

In application, this method depends on the passage of a current of ambient air contaning CO over a heated (150 to 160°C) tube after the air stream has been purified by passing through a series of interference scrubbers. Activated carbon is used to remove hydrocarbon interferences, while a packed column of glass beads on which chromic acid has been deposited acts to remove ozone and nitrogen oxides. A tube containing packed Ascarite can be used to separate these two scrubbers if the CO_2 produced from this reaction is to be measured after passage of CO over I_2O_5. It is important to condition this assembly for a period of several hours to several days by passage of pure nitrogen at a flow of approximately 50 ml/min with the I_2O_5 held at a temperature of 150 to 160°C.

The actual measuring procedure involves injection of an ambient air sample into the conditioned oxidizing train using either a gas pipette or, in the case of a more concentrated sample, a hypodermic needle. The resulting CO_2 is absorbed on a tared Ascarite tower, whose gain in weight can be used directly to calculate the milligrams of CO present in the original sample volume by

$$(\text{weight increase of Ascarite tower}) \times \frac{28}{44} = \text{mg CO}$$

It must assumed that enough sample was attained to yield a weighable amount of CO_2 (at least 10 mg) and that any I_2 is removed upstream of the Ascarite tower by use of a small (10 cm³) column packed with finely divided copper.

In cases where only very small field samples (10 to 100 ml) of actual gas volume have been obtained, it is better to titrate the iodine (I_2) that is formed according to the above reaction. The liberated I_2 is collected using carbon tetrachloride and can be determined colorimetrically or titrimetrically by dissolving in 2 to 3% potassium iodide and titrating with standard 0.001 sodium thiosulfate solution or 0.01 N arsenite solution to a starch end point.

An excellent review of the fundamentals involving the I_2O_5 method is provided by Adams and Simmons.[22] Also, a modified procedure suggested by

Hersh and Sambucetti employs determination of the liberated iodine by coulometry.[23]

The described procedure can be used for in-laboratory analysis of CO where no infrared or gas chromatographic instrumentation is available. However, CO should be analyzed using the automated methods described in Chapter 4 if at all possible.

Manual Colorimetric Method[24]

When an alkaline solution of the silver salt of p-sulfaminobenzoic acid is allowed to react with CO, a colloidal suspension of silver is formed, whose absorption maximum at 425 mμ can be determined by spectrophotometry.

The low range below 20 ppm can be analyzed to the 2 ppm minium detectable limit using a sample volume of 125 ml. Concentrations in the range of 400 to 1800 ppm can be determined by the same method using the less-sensitive absorbance peak at 600 mμ.

Where this reaction depends on the weak reducing properties of CO, other reducing gases (e.g., H_2S or olefins) can interfere through reduction mechanisms on the precipitation of Hg_2S. Where such interferences are known to exist, passing the air sample through a filter of mercuric sulfate deposited on silica gel will remove most common sulfide or nonsulfide reducing agents.

Reagents

1. 0.1 p-sulfaminobenzoic acid solution is prepared by dissolving 2.0 g of the reagent in 100 ml of a 0.1 M NaOH solution.
2. 0.1 M AgNO$_3$ is prepared from cp silver nitrate.
3. 1.0 M NaOH is prepared from cp sodium hydroxide.

Procedure

Mix 20 ml of p-sulfaminobenzoic acid reagent with 20 ml of 0.1 M silver nitrate. To this mixture, add 10 ml of 1.0 M NaOH a little at a time with mixing until a clear colorless solution results.

Place 10 ml of this absorbing solution into a flask that is fitted with a ground-glass stopcock, and evacuate the flask through the stopcock until the absorbing solution begins to boil. Admit a measured volume of sample air into the flask by sleeving the sample-collecting pipette to the stopcock of the evacuated flask or by injecting a sample by gas pipette as the stopcock is opened. In case of suspected interferences, admit the sample through the HgSO$_4$ silica gel filter. Return the flask to atmospheric pressure by opening the stopcock to pure air, and shake the flask on an automatic shaker for a period of 2 hr ±5 min. Standards prepared from known volumes of CO are prepared in the same manner, making sure that standards and samples have the same color development time during shaking.

Calculation and Calibration

A standard curve is drawn using known dilutions of pure CO, which are prepared according to the method described in Chapter 5. Gastight microliter syringes are used to transfer aliquots of CO to the evacuated absorption flask. Consequently, the number of microliters of CO can be derived directly from the standard curve, where ppm of CO = (μl of CO)/liter of sample air.

This method has been adopted as a tentative method for manual analysis of CO by the Intersociety Committee on *Methods of Air Sampling and Analysis,* January, 1970.[25,26]

Other Chemical Methods for Determining CO

Other laboratory methods for CO include combustion, which involves burning with oxygen or passage of the sample in air over cupric oxide at 280 to 300°C. The resulting CO_2 can then be measured directly by Ascarite absorption or indirectly by the change in conductivity of a calcium hydroxide solution.[27] Also, a gravimetric method involving the reduction of HgO vapor reaction with CO at 175 to 200°C has been researched.

As a final absorptimetric method, an adaptation of the Hopcalite method (described in Chapter 4) can be used as a batch process to produce CO_2, which is absorbed and weighed.[28] Briefly, the method involves drawing of CO-containing air over Hopcalite, a catalyst consisting of MnO_2 and CuO. Single samples of ambient air can thus be conveniently analyzed for CO to levels above 20 ppm.

More recently, Lysyj[29] has proposed the use of silver permanganate ($AgMnO_4$) as an oxidation catalyst for CO, which is absorbed as CO_2 from single samples in the manner of the Hopcalite and I_2O_5 methods.

Instrumental Method Using Infrared Spectrophotometry

When infrared spectrophotometry is available, an in-laboratory method can be employed to determine CO in a single air sample using the 10-m infrared gas cell.

Of the two CO absorption maxima at 4.67 and 4.72 μ, the 4.67 peak (2143 cm^{-1}) with a slow chart-speed setting is used to measure concentrations of contaminant in excess of 10 ppm.

Although the selective absorption peak effectively screens out most interfering molecules, gases that absorb in the 4.6 to 4.7 μ region include cyanogen, H_2S, nitrous oxide, some olefins, nitrosyl chloride, some aldehydes, and acetylene. Of these, H_2S constitutes the most notable interference because of its normal presence in an industrial atmosphere. Carbon dioxide at levels up to 3% has not been found to interfere.[30]

The method requires an infrared spectrophotometer having a 2 to 15 μ range with a wavelength accuracy of 0.015 μ.

Procedure

Using the flow scheme shown in Figure 2, the 10-m gas cell is connected to a manometer through a T-fitting, and a vacuum pump is used to evacuate the cell to a pressure of approximately 1.0 mm Hg. The sample containing CO is then introduced into the gas cell through a calcium chloride drying tube until pressure equilibrium is achieved, with this pressure noted on the manometer. The gas cell can then be equilibrated to atmospheric pressure with CO-free air. The peak area at 4.67 μ is then used to determine the number of microliters of CO present.

Calculation

To determine the parts per million concentration of CO, the exact volume of the 10-m cell must be known. This volume can be determined by evacuation of

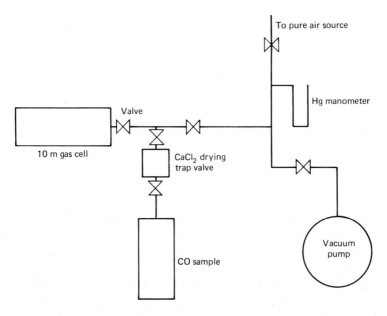

Figure 2. Introduction of gas samples into infrared cells. Schematic diagram showing arrangement of equipment for analysis of CO by the manual infrared method. [Reprinted by permission from American Public Health Association. *Health Lab. Sci.*, **7**(1), 79 (1970)].

the cell with readmission of air through a wet test meter to determine actual volume in cubic feet.

Manual Hopcalite Method

The catalytic CO oxidation, or so-called *Hopcalite* method, for CO is based on a method for removal of CO from air and dates back to one of several methods studied for extracting CO from respirable air during World War II.[31] The method was later studied and revised to yield reaction conditions under which CO is quantitatively oxidized to CO_2. Quantitative detection is based on either the volume or weight of CO_2 produced or the increase in temperature of the exit airstream because of the heat of oxidation, which is proportional to the CO concentration.[32]

In this method, the actual oxidation catalyst, Hopcalite, is described as consisting of a mixed bed of more than two oxides. However, for purposes of analytical application, the definition of the catalyst stoichiometry as MnO_2 plus CuO is sufficient.

Although the method is normally applied to atmospheres containing in excess of 50 ppm CO, some commercially available instruments quote an accuracy of ±5 ppm for a display scale of 0 to 500 ppm total span.

Positive interference may be observed from such reducing agents as low-molecular-weight aldehydes and ketones. Methane, ethane, and propane do not interfere below 150°C, and the aromatic hydrocarbons do not interfere below 200°C.[33] Consequently, the effect of these possible interferences can be reduced by selection of an appropriate oxidation temperature.

The in-laboratory manual method would normally employ the catalyst as a medium in which to oxidize CO and would measure the resulting CO_2 rather than the temperature change of the exit products.

Procedure

A particulate filter and a dryer unit are usually employed to pretreat the sample air. The sample air is then drawn through a controlled-temperature cell containing Hopcalite by means of a diaphragm pump that is fitted with a simple mechanism for sample flow regulation to ensure a uniform and controlled sampling rate. The oxidation cell usually operates at a controlled temperature of approximately 100°C, as regulated by a Variac, to minimize interferences from other organic compounds.

Gravimetric measurement is accomplished simply by introduction of a measured volume of ambient air (by means of a gas pipette) into the sample line ahead of the filter-dryer assembly, as shown in Figure 3. If the thermal method of measurement is employed, sample introduction can be by batch, intermittent, or continuous.

The catalyst cell is made up of two Hopcalite beds—one active and the other

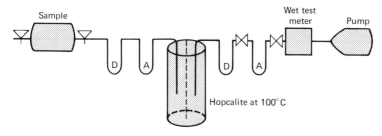

Figure 3. Hopcalite conversion train. Sampling assembly for gravimetric measurement of CO using hopcalite. A—Ascarite trap; D—Drierite trap.

inactive—with the inactive bed serving as a reference path. A single thermistor placed in each Hopcalite bed is connected one to each leg of a Wheatstone bridge assembly. As a result, although CO passes through both cells, it is oxidized only in the active cell, thereby producing a cell temperature differential that is measured by the bridge.

Calibration

Calibration of the instrument is accomplished by using a dilution volume of known CO. Thermal response is a linear function of the CO concentration.[34]

Calculation

Parts per million of CO can be read directly from a calibration curve in volts when the thermal method is used to monitor the temperature of exit gas.

Gravimetrically, weighed milligram amounts of CO_2 can be converted to ppm of CO by the formula

$$\text{CO, ppm (volume at 20°C)} = \frac{517.6 \times \text{mg } CO_2}{\text{sample volume in liters}}$$

Preparation of Standard CO Mixtures

In review of the foregoing laboratory methods for determining ambient CO, the reader will see the need for producing a reliable standard CO reference dilution. Although several metal carbonyls are available as possible chemical standards, considerations involved in solving both their decomposition stoichiometry and determining their expected shelflife have limited the widespread acceptance of these compounds as analytical standards. For this reason, a tentative method for the preparation of standard CO mixtures has been researched by the Intersociety Committee for *Methods of Ambient Air Sampling and Analysis*.

This preparative method is based on the principle that aliquots of cp CO obtained from cylinder sources can be volumetrically diluted with CO-free air and subsequently calibrated by oxidation to CO_2, using the catalyst Hopcalite. The CO_2 is then measured gravimetrically by absorption on Ascarite.[35,36]

For replicate analyses to yield precision of $\pm 2.5\%$, it was found necessary to produce a minimum weighable amount of 10 mg of CO_2 to achieve good reproducible accuracy. Consequently, sufficient sample volume must be collected to ensure adequate CO. However, practical considerations often limit the total sample volume to approximately 60 liters. At a flow of ½ liter/min, this volume would require 2 hr for completion of the analysis. Such a 60-liter sample at a concentration of 100 ppm of pure CO would produce 10.8 mg of CO_2 (10 ppm of CO in air at 25°C represents 11.45 mg/m^3 at standard pressure[37]).

Procedure

An evacuated 34-liter stainless steel tank is used as a mixing chamber for the standard CO concentrations. A standard airtight Luerlok gas syringe equipped with a 24-gauge needle is used to introduce aliquots of between 50 and 500 ml of cp CO into the mixing chamber through a rubber septum, which can be prepared from a serum bottle cap. The tank is then filled with CO and CO_2-free air. This air can be prepared by passing dried room air over Hopcalite at 100°C and then over Drierite, followed by Ascarite. Dilution of pure CO can be allowed to continue until measured atmospheric pressure is achieved or until some desired pressure is reached where the actual dilution volume will be several times greater than the volume of the tank because of increased pressure.

The mixture usually should be allowed to stand for 4 to 5 hr to achieve thorough mixing before a standard is withdrawn. Also, care should be exercised in determining that any transfer syringes are airtight and that the volumes of the syringes as well as that of the 34-liter tank are accurate.

Removal of aliquots from this standard mixture can be accomplished by syringe, or a pump can be used to draw a measured volume continuously. A wet test meter at the end of the system provides an accurate measure of the volume of air passed.

Calibration

Calibration of these standard volumes has shown that standards in the range of as low as a few parts per million can be prepared to an error of less than 5% where the absolute analysis for calibration of these standards is accomplished using the Hopcalite method shown in Figure 3.

In this method, a Drierite trap is inserted to remove moisture ahead of an

Ascarite trap that serves to remove any background CO_2 before the standard mixture enters the Hopcalite bed, which is heated to 100°C. A second Drierite trap is placed downstream of the catalyst to remove any moisture produced in the oxidation, and a tared Ascarite trap is used to collect CO_2, from which CO is calculated.

All traps should be borosilicate glass-type with a recommended ID of 7 to 10 mm. The traps should be approximately 15 cm in length for proper contact-retention. Glass wool should be used to plug each end, and ground-glass joints are recommended for ease of assembly and disassembly.

A purge of the system is required before any analysis. Here, after removal of the tared Ascarite trap and with the catalyst at 100°C, the oxidizing train is purged with pure nitrogen at ½ liter/min for several hours.

Calculation

Where 1 mg of CO_2 at 25°C and standard pressure is equivalent to 555.54 ppm CO

$$\text{CO, ppm} = \frac{\text{mg } CO_2 \times 555.54}{\text{sample volume in liters}}$$

CARBON DIOXIDE (CO_2)

In a normal uncontaminated atmosphere, the concentration of CO_2 varies from 300 to 350 ppm. Carbon dioxide is not considered hazardous at this level, and normal daily fluctuations at this level are not of concern. Consequently, measurements of CO_2 are made less frequently, and methods for its determination in ambient air have been considered of secondary importance.

Procedure

If particular circumstances dictate the need for CO_2 measurement, the air sample can be passed through a fritted glass bubbler containing 25 to 50 ml of 0.01 M barium hydroxide, which is 6% in barium chloride and contains 0.2 to 0.4% n-butanol.

When sampling is completed, the contents of the bubbler containing excess barium hydroxide are titrated with a standard 0.1 N oxalic acid solution using five drops of a mixed end-point indicator that consists of 1.0% phenolphthalein and 0.5% thymolphthalein in alcohol.[38]

Calculation

$$0.01 \text{ meq acid or base} = 0.22 \text{ mg } CO_2$$

$$\text{parts per million } CO_2 \text{ (vol at } 25°C, 760 \text{ mm)} = \frac{\text{mg } CO_2 \times 873}{\text{sample volume in liters}}$$

Precautions

Care should be taken to prevent contact of the absorbing solution with ambient air before sampling begins, as well as during the titration. Such contact is minimized by carrying out the titration directly in the absorber through a sleeve coupling of flexible household food wrap while the solution itself is agitated by means of a stream of CO_2-free nitrogen that is passed through the fritted bubbler.

Other Methods of Measurement

Finally, a number of indicator tube methods are employed for the detection and determination of CO_2 in concentrations varying from 10 to 100 ppm to an accuracy of 5 to 10%. These length-of-stain-indicator tubes commonly employ potassium palladosulfite,[39] I_2O_5,[40] or palladous silicomolybdate.[41]

Sulfur Dioxide (SO_2)

Sulfur dioxide and sulfur trioxide (SO_3), along with their corresponding acids and particulate salts, are common pollutants in an urban industrial atmosphere. These gases arise chiefly from the combustion of fossil fuels whose sulfur content as metal and organic sulfides varies from less than 1% to more than 5%. Combustion of 3% sulfur coal in power plants produces an effluent that averages 1000 to 2000 ppm of SO_2 and as much as 20 to 40 ppm SO_3.

Sulfur dioxide is not flammable and can be detected by its characteristic "burning match" odor at concentrations somewhat below 0.2 ppm. Above 0.3 ppm, it is subject to detection by taste. At levels above 1 ppm, it produces a biting acrid sensation to the nose.[42]

Sulfur dioxide is not ordinarily stable in the atmosphere, being oxidized to SO_3 by a variety of catalytic[43] as well as suspected noncatalytic mechanisms.[44] Catalysts include oxides of nitrogen and compounds of iron and manganese. Noncatalytic mechanisms depend on the presence of sunlight or ozone.[45]

Sources of SO_2

Sulfur dioxide is normally present in the atmosphere is concentrations varying from 0.02 to 0.1 ppm. By far, the largest single source of this contaminant gas is the generation of electric power. Other combustion sources (e.g., refineries, smelting, coking, and chemical industry) provide degrees of total contribution according to data published in *Air Quality Criteria for Sulfur Oxides* and summarized in Table 1.

Probably the most important consideration in evaluating the atmospheric profile of sulfur oxides is the meteorology of their dispersion in the atmosphere. The past 20 to 40 years have witnessed "control" of sulfur oxides from power generation by consolidation into large point sources. For example, single stacks use stack elevation as the chief controlling feature in dispersal of the high ppm effluent by dilution at heights of 400 to 600 ft. Therefore, it is not unusual for large urban centers to experience generally high background concentrations of SO_2, because of this uniform dispersion effect.

Of interest to the air analyst is the fact that, although high stacks tend to reduce average ground level concentration of SO_2, episodes of low winds and/or thermal inversion result in fumigation conditions that can affect even greater areas of population than would low-level random SO_2 point sources with the same meteorological conditions. Consequently, air sampling for SO_2 in urban environments should not be limited to a few analyzers placed at intervals from one or more major power facilities. Instead, air sampling should be a collective product of many analyzers distributed throughout the community ac-

Table 1 Atmospheric SO_2 Emissions by Source[46]

Process	Tons of SO_2	Percentage of Total SO_2 Emissions
Power generation	11,925,000	41.6
Other combustion	4,700,000	16.6
Combustion of oil and other petroleum products	5,604,000	19.6
Refinery operations	1,583,000	5.5
Ore smelting	3,500,000	12.2
Coking and sulfuric acid manufacture	1,050,000	3.7
Incineration and other sources	200,000	0.8

cording to either equal geographic areas (New York) or equal population densities (Detroit).

Other Sulfur Oxides

Oxidation of sulfur is known to produce a variety of gaseous sulfur oxide compounds, such as SO_3, S_2O_3, and S_2O_7. The main product is probably SO_3, which yields sulfuric acid mist upon hydrolysis with rain water or water vapor. Oxidation of SO_2 by ozone has been reported,[48] with the production of S_2O_7.

Sulfates that result from these reactions can account for as much as 10 to 20% of the total particulate found in the suspended dust collected from air sampling in urban areas.

When combined total air samples have been collected to determine the ratio of SO_2 to total sulfur compounds, the average SO_2/SO_x ratio was found to be highest at points of highest SO_2 concentration. This ratio was found to decrease from approximately 0.95 at the end of 1 hr to 0.65 at the end of 12 hr.[47] Although debatable points arise in connection with the actual stability of the atmosphere and uniformity in output of the source, the tendency toward oxidation of SO_2 is at least indicated by the increases of reaction products as SO_x.

Among other aspects of the SO_2 oxidation is the effect on visibility of atmospheric sulfate as suspended particulate. This particulate matter is recognized as an important source of haze,[48] and sulfate particle size distribution has been found to be closely related to relative humidity.[49] In addition, where the average mass median diameter of particulate was found to fall between 0.3 to 0.4 μ in sampling in Chicago and in Cincinnati,[50] several investigations have shown that 80% or more of atmospheric sulfate is in the particle size range below 2 μ in diameter, with most of it being in the 0.2 to 0.9 μ range.[49,50]

Methods of Measurement

The most common manual methods for measuring ambient SO_2 are based on iodimetry, turbidimetry, or colorimetry. A number of less-common manual methods using conductometric, potentiometric, or coulometric end-point detection is also employed, but their application is more appropriate to continuous air monitoring as described in Chapter 5.

Of the methods most often employed, the manual colorimetric method of West and Gaeke is considered by most air pollution chemists as the standard or reference method.

Colorimetric Method for SO_2 (Modified West Gaeke)[51]

This method, which is intended for both field and laboratory use, is selective and quite sensitive for SO_2. Application depends on drawing of ambient air through 0.1 M sodium tetrachloromercurate, which is prepared conveniently from aqueous soluble mercuric chloride and sodium chloride. Sulfur dioxide is absorbed by this reagent to form the stable and nonvolatile dichlorosulfitomercurate(II) complex $(HgCl_2SO_3)^{-2}$, which then behaves effectively as "fixed" SO_3^{2-} in solution. Subsequent color-forming reaction of this sulfite ion with formaldehyde and bleached pararosaniline in acid solution yields the reddish purple pararosaniline sulfonic acid. The color intensity of this acid at 548 mμ and pH 1.6 $\pm$ 0.1 is proportional to the concentration of absorbed SO_2 over the concentration range of 0.002 to 5 ppm.

Recent updating of the method has resulted in an improved standard form of the dye, pararosaniline, to achieve greater reproducibility of measurements and flexibility of pH range in color formation.[52]

Reagents (ACS Reagent Grade or Equivalent)

i Absorbing Reagent Absorbing reagent 0.04 M potassium (or sodium) tetrachloromercurate is prepared by dissolving 10.8 g of mercuric chloride together with 5.96 g of potassium chloride (or 4.68 g of sodium chloride) and 0.066 g of the disodium salt of EDTA in distilled water to a total volume of 1 liter. (*Note*: Mercuric chloride is highly poisonous. If spilled on skin, promptly flush off with water.) The pH of this solution should be checked, and the solution rejected if the pH is below 5.2. This reagent is usually stable for up to 6 months, but it should be discarded if formation of a precipitate is noted.

ii Pararosanline Stock Solution 0.2% Using purified 99.0% pararosaline, dissolve 200 mg of the dye in 100 ml of 1 N HCL. This solution should be assayed and repurified if pararosaline other than purified 99.0% is used. The assay and purification are described by the Intersociety Committee, *Methods for Ambient Air Sampling and Analysis*.[52]

iii Pararosanline Reagent Using a 250-ml volumetric flask, dilute 20 ml of pararosaline stock solution to volume with distilled water.

iv Formaldehyde 0.2% Dilute 5.0 ml of 40% formaldehyde to 1 liter of aqueous solution. Solution is stable for approximately 1 to 2 days.

v Standard Sulfite Solution Dissolve 0.400 g of sodium sulfite or 0.300 g of sodium metabisulfite in 500 liters of recently boiled or freshly double distilled water. This solution, which contains approximately 320 to 400 μg SO_2/ml, should then be standardized by adding an excess of iodine (50 ml of 0.01 N iodine solution per 25 ml of standard sulfite solution) and back-titrating the excess iodine with dichromate-standardized 0.1 N sodium thiosulfate solution to a starch end point.

The actual SO₂ equivalent is then calculated from the relationship

$$SO_2, \mu g/ml = \frac{32(A - B) \times 10^3}{25}$$

where A is the number of milliliters required for sample, B is the number of milliliters required for titration of a 25-ml distilled water blank.

vi Dilute Sulfite Solution Once the sulfite solution has been standardized, pipette 2.0 ml of the standard sulfite solution into a 100-ml volumetric flask and dilute to volume with 0.04 M potassium tetrachloromercurate solution. This solution remains stable for approximately 30 days if refrigerated.

PROCEDURE. Into a midget impinger, add 10 ml of 0.04 M tetrachloromercurate absorbing solution. (If a Greenberg-Smith impinger is used, add 100 ml of TCM reagent.)

Connect the impinger to a suitable glass or Teflon probe and, using a rotameter, adjust the flow of sample air to 0.5 to 2.5 liter/min. The impinger containing absorbing reagent should be protected from direct sunlight to prevent oxidation breakdown of complex; in the field, this can be achieved by wrapping the impinger with aluminum foil. Once collected, it is advisable to refrigerate the sample if longer than 1 day's time will elapse before analysis. Centrifuge to separate any precipitate that may form on standing.

Rinse the contents of the impinger into a 25-ml volumetric flask, using no more than 5 ml of distilled water. Add 1 ml of 0.6% aqueous sulfamic acid and allow to stand for at least 10 min. Add exactly 2.0 ml of 0.2% formaldehyde solution, followed by 5.0 ml of pararosaniline reagent. Dilute to volume (25 ml) with distilled water. If several samples are to be analyzed at the same time, this final dilution should be made within a time interval of 1 to 2 min from the first to last sample. After exactly 30 min has elapsed, the absorbances of the samples and a reagent blank are determined, in the order of dilution, at 5.48 mμ.

When absorbance is greater than 1.0, dilute the sample with an equal portion of reagent blank and read after allowing to stand for 3 to 5 min. Highly colored solutions can be diluted to a ratio of 6:1, if necessary.

CALIBRATION. Pipette 0, 1, 2, ... ml aliquots of standard sulfite solution into a series of 25-ml volumetric flasks and dilute to about 10 ml with 0.04 M TCM. Next, add 1.0 ml of 0.6% sulfamic acid and allow to react for no less than 10 min. Add 20 ml of 0.2% formaldehyde and 5.0 ml of pararosaniline. Dilute all samples to volume with distilled water. Allow to stand 30 min and read absorbance as described in *Colorimetric Analysis*.

Plot the absorbance of each solution versus micrograms of SO₂. The calibration factor so obtained is defined as the reciprocal of the slope of the line.

CALCULATION. The volume concentration of ambient SO_2 is then given by

$$SO_2, \text{ppm} = \frac{(A - A_0)\, 0.382B}{V},$$

where A is sample absorbance, B is the calibration factor, A_0 is blank absorbance, V is the sample volume in liters corrected to 25°C and 760 mm, using the gas law ($pV = nRT$).

INTERFERENCES. Negative interference is observed from iron and other heavy metals whose salts may oxidize the SO_2 prior to complex formation. EDTA is often added to sequester such metal interferences. Nitrogen dioxide interference can be eliminated by adding a small amount of sulfamic acid to destroy the nitrite ion prior to color formation.

Other Manual Methods for Determining Atmospheric SO_2

IODIMETRY. Using an impinger containing 10 ml of dilute $1.7 \times 10^{-5}\, N$ standard iodine solution, sample air is admitted by bubbling. Absorbance at 450 mμ versus a reagent blank is used to determine the concentration of ambient SO_2, once the absorbing system is calibrated using a known concentration of SO_2.[53]

TITRIMETRY. Ambient SO_2 is absorbed in 0.03 N hydrogen peroxide at pH 5. Sulfur dioxide so absorbed is quickly oxidized to sulfuric acid, which is titrated with standard alkali.[54] Where few interferences, such as NO_2 or O_3, are suspected at concentrations equal to SO_2, this method yields good accuracy and reproducibility comparable to the pararosaniline method, although the latter titrimetric method is much simpler.

This method also has been adapted for very low (0.01 ppm) concentrations of SO_2 using turbidimetry to disperse the barium sulfate solution resulting from addition of crystalline barium chloride to the sulfate formed from acid hydrogen peroxide absorption of SO_2. A glycerol–alcohol solution is used to prevent settling of the precipitate, whose end-point turbidity is determined at 450 mμ using a standard spectrophotometer.

Lead Peroxide (Reactive Surface Sampling Method)

The reaction of SO_2 with lead peroxide ($PbO_2 + SO_2 \rightarrow PbSO_4$) produces a solid, stable lead sulfate. This method of determining SO_2 has experienced widespread use in England since its proposal by Wilsdon and McConnell in 1934.[55]

The object of this method is to arrive at a reproducible index of SO_2 as a measure of its effects on average reactive surfaces (e.g., paint or mortar), with some allowance made for the inherent physical limitations of the method and known interferences. Hydrogen sulfide and sulfuric acid aerosols can interfere by contributing to the "sulfation index,"[56] while the deactivation of the surface that can occur because of poor preparation techniques may lead to low index results.

In principle, the method involves only the preparation of an active reagent paste containing lead peroxide homogenized in gum tragacanth solution. The paste can be either (1) mixed with fiberglass filter flakes and poured into a shallow dish to form a "plate," or (2) used to coat cotton gauze that is wrapped around a porcelain cylinder to form a "candle." Although a correlation factor is required to compare the activity of plates with candles, either of these mountings is suitable for obtaining this air sampling index.

After exposure, the reactive surface is separated from the dish or cylinder support material, and sodium carbonate is used to convert insoluble lead sulfate to soluble lead carbonate. The remaining lead peroxide can then be separated by filtration and sulfate determined according to a choice of methods, such as gravimetry, turbidimetry, or titration.

This reaction rate of SO_2 with lead dioxide, which is first order on a fresh surface assuming SO_2 concentrations no greater than 700 ppm, is dependent on both wind speed and temperature. It is well to note that relatively long exposure times (more than 28 days at 0.02 ppm SO_2 or 14 days at 0.2 ppm SO_2) may lead to saturation of the surface. In this event, the remaining available surface area, rather than level of SO_2, becomes the rate-determining variable.

PREPARATION OF REACTIVE SURFACE AND EXPOSURE OF CYLINDERS. Fill a 48-mm ID plastic petri dish with active mixture, which is prepared in the following manner. Using a laboratory blender, blend 0.7 g of gum tragacanth with 7 g of Gelman-Type A glass-fiber filters and 112 g of PbO_2 in 700 ml of distilled water. (*Note*: A better emulsion can be achieved by mixing the gum tragacanth with a few drops of dioxane before adding to water. This also serves to etch the surface plastic of the petri dish for better retention of the active material.)

Approximately 10 ml of this mixture is used to fill the petri dishes, which are then oven-dried at 60°C. When the dishes are dried, they can be covered with plastic lids, thereby serving as storage containers. Specific activity of these plates compared to sulfation candles was found to be slightly higher by a factor of 1.19,[57] although others[58] suggest higher sulfation rates for the cylinder than for the plate over equal times of exposure.

ANALYSIS OF EXPOSED LEAD PEROXIDE. (*Caution*. Lead peroxide is highly toxic; use rubber gloves.) The contents of an exposed lead peroxide plate are scraped or snapped from the petri dish into a 250-ml beaker.

Fifty milliliters of 5% sodium carbonate solution are then added, and the solution is heated for 4 hr to dissolve the lead sulfate. The contents of the beaker are then filtered through Whatman No. 12 filter paper. Four to six drops of methyl orange indicator are added, and the solution is neutralized to an actual pH of 4 to 5 (use pH meter).

The acidified filtrate is then diluted to 100 ml using a volumetric flask. A 10-ml aliquot is withdrawn and diluted to 20 ml. To this solution, 0.1 g of SulfaVer powder (Hach Chemical Co., Ames, Iowa) is added, and the resulting turbidity is measured in a spectrophotometer at 450 mμ.

A blank can be prepared from an unexposed sulfation plate and its value subtracted from the turbidity of the samples measured.

A rapid combustion method for liberating SO_2 from exposed cylinders with subsequent iodometric titration has also been proposed by Vijan.[59]

CALCULATION. Results are reported as sulfation rate in milligrams of SO_3 (calculated from $BaSO_4$) per 100 cm^2 of exposed surface per month. The area of the plate or cylinder must be known or calculated as exactly as possible.

For cylinders, the gravimetric determination yields

$$(\text{mg } SO_3)/(100 \text{ cm}^2)(\text{month}) = 1029 \frac{W_1 - W_2}{AT}$$

where W_1 is the weight of $BaSO_4$ (milligrams) or an exposed cylinder; W_2 is the number of milligrams of blank unexposed cylinder; A = area of exposed reactive surface (square centimeters); and T = time (number of actual days exposure).

Sulfation rates in the United States have been recorded to vary from 0.05 to 8.0 (mg of SO_3)/(100 cm^2)(day).

Although it is inadvisable to attempt conversion of sulfation rate into parts per million of ambient SO_2, Huey suggests a conversion factor of parts per million (average per month) = sulfation rate (month) × 0.03.[57] Others suggest correlation factors of 0.06 or 0.09.[58]

PRECISION AND ACCURACY. Precision and accuracy have been found by the author and others to be largely a function of reproducibility of preparation techniques. With normal precipitation techniques, Thomas and Davidson report no greater than 10% error observed for triplicate cylinders exposed at four sampling sites,[60] and Keagy et al.[61] suggest that annual average sulfation rates can be estimated to a 10% accuracy within a confidence interval of 95%.

Plates can be exposed for as short a time period as one day with detectable sulfation in average atmospheres.[62]

SULFURIC ACID MIST

Due to its high boiling point (338°C), sulfuric acid exists in the atmosphere as particulate.[63] Hence it is properly referred to as a mist.

Sources of Acid Mist[64]

The production of sulfuric acid mist from sulfuric acid manufacturing as well as general chemical industry has resulted in local episodes involving this particulate.[65] A study of gas-phase oxidation reactions by Bufalini[66] has shown that, in the presence of water vapor and suitable metal salt containing suspended particles, SO_2 is capable of combining with NO (or air oxygen) to form SO_4^{2-} in a series of photolytically induced reactions. Depending on the presence of alkali metal ions, the resulting sulfate may or may not increase the total acidity of the atmosphere. Consequently, either of two approaches can be used to determine the ambient level of sulfate.

Determination of Acid Mist

Particulate sulfuric acid mist can be collected by passing ambient air through a filter paper that has been repeatedly distilled water leached to remove residual acidity. Such a desiccated paper can be placed in a filter holder or high-volume support frame (see Chapter 6) and a known volume of 60 to 100 ft³ of air passed through the filter. Subsequent titration of the acidity using 0.002 N sodium hydroxide is corrected by means of an unexposed blank paper, and the acidity as H_2SO_4 is expressed in parts per million as

$$\text{ppm } H_2SO_4 = \frac{(\text{ml NaOH}) \times N \times 22.41 \times 10^3}{2 \times 28.32 \times V}$$

where V equals the volume of sample air, in cubic feet, at standard conditions.[67]

Determination of Sulfate Aerosol

Airborne particulate obtained by the above sampling method by high-volume filtration or by means of A. W. Andersen's sampler (Chapter 5) can be analyzed directly for aerosol sulfate by turbidmetry,[68] nephelometry,[69] or the barium chloranilate colorimetric method.[70]

Summary and Comparison of Sulfur Oxide Methods

In Europe, adaptations of a method employing titration of sulfuric acid produced on reaction of SO_2 with hydrogen peroxide is used for both manual and continuous sampling. In the United States, the colorimetric pararosaniline method is used for manual SO_2 determination, and the coulometric method has been adopted by many state and municipal agencies for continuous air monitoring.

Since these methods do not always measure the same form of the air contaminant, it is not surprising that they do not always correlate well with one another, although it can be inferred from the general specificity of the method that the greater portion of any response is due to SO_2.

In general, acid interferences, such as hydrochloric acid, contribute to the ambient level obtained by the hydrogen peroxide method.[70] A recent study[71] has shown that the overall effect of several interferences tends to produce higher results by the acid peroxide method when conductivity end-point detection is employed.

HYDROGEN SULFIDE (H_2S)

Sources of H_2S

Ordinarily, air pollution by H_2S is not a widespread problem. However, local sources [e.g., Kraft paper processes, sewage plants, coke ovens, and steel making (slag quench)] in an urban industrial atmosphere can produce strong isolated fumigation by this colorless, highly odorous gas.

Aside from industrial sources, H_2S occurs naturally as a product of the decomposition of proteinaceous matter through bacterial action, most commonly in swamp or stagnant water. In addition, H_2S is found as a constituent of natural gas, volcanic gases, and sulfur springs.

Odor Threshold of H_2S

The odor threshold for H_2S produced according to the recognition threshold method (see Chapter 7) is 0.00047 ppm.[72] However, according to the study of Leonardos et al.,[73] the odor threshold is 0.0047 ppm because of the effect of water vapor and the impurities present in sodium sulfide crystal. It is interesting to note that Leonardos's study was prepared with the intent of superseding previous odor threshold values for H_2S. The early literature shows only scattered threshold data, as shown in Table 2. This may account for an actual human response gradient as reported by the U.S. Department of Health, Education and Welfare.[74]

It was found that the odor of H_2S is "distinct" at 0.350 ppm, offensive and moderately intense at 2.80 ppm to 5.6 ppm, and strong but not intolerable at 21 ppm to 35 ppm.[81]

Table 2 Odor Threshold Data for H_2S

Odor Threshold (ppm)	Date	Reference
0.0065–0.0324	1965	Adams and Young[75]
0.002	1966	Larson[76]
0.005	1968	Wilby[77]
0.008–0.020	1959	Fyn-Djui[78]
0.7	1939	Dalla Valle[79]
0.7	1949	McCord and Witheridge[80]

The odor is not as pungent at 226 ppm as at lower levels because of paralysis of the olfactory nerves.[82] Although threshold concentrations are based on "initial inhalations" using a fresh subject, it is noteworthy that repeated inhalation does cause rapid fatigue of the olfactory sense; thus, at concentrations above 7840 ppm, where there is very little odor stimulus, death can occur without olfactory warning.[83] Consequently, due to the dangerous effect (loss of sense of smell) of prolonged exposure to moderate levels of this gas, it is essential that protection from the harmful effects of H_2S be ensured through establishment of a low average urban concentration of H_2S at between 0.001 and 0.009 ppm.

Methods for H_2S Determination

The low odor threshold of H_2S in ambient air makes its chemical detection at or near this odor threshold all the more necessary for purposes of monitoring-recording. Although many methods have been reported in the literature, the most sensitive method for air pollution monitoring is the colorimetric method of Jacobs,[84] which has been proposed as a tentative method for H_2S in ambient air by the Intersociety Committee for Methods of Air Sampling and Analysis.

Colorimetric Methylene Blue Method (Jacobs)

In this method, H_2S is absorbed as basic cadmium sulfide (CdS) when ambient air is passed through a suspension of cadmium hydroxide in an alkaline solution at a sampling rate of 1 cfm. The absorbed sulfide ion is redissolved and reacted with a mixture of ferric chloride and N,N-dimethyl-p-phenylenediamine in strongly acid solution to produce methylene blue, which is determined spectrophotometrically. Stractan 10 is added to the suspension of cadmium hydroxide to prevent photodecomposition of the CdS during collection.

Precision and Accuracy

The sensitivity of the method depends on the collection efficiency and sampling time. Collection efficiency is a function of the design of the scrubber. Sampling time can be as short as 10 to 15 min, but ½ hr is recommended at a level of 1 ppb H_2S.

Reagents (ACS Analytical Grade; Solutions Should Be Kept Refrigerated.)

i Acid Amine Stock Solution   Fifty milliliters of concentrated sulfuric acid are dissolved carefully in 30 ml of distilled water and allowed to cool before the addition of 12.0 g of N,N-dimethyl-p-phenylenediamine dihydrochloride.

ii Acid Amine Test Solution   Twenty-five milliliters of the acid amine stock solution are diluted 1:1 with sulfuric acid.

iii Ammonium Phosphate Solution   Four hundred grams of diammonium phosphate are dissolved in water and diluted to 1 liter in a volumetric flask.

iv Ferric Chloride Solution   One hundred grams of $FeCl_3 \cdot 6H_2O$ are dissolved in water and diluted to 100 ml.

v Absorbing Solution   Into separate portions of water, dissolve 4.3 g of cadmium sulfide octahydrate and 0.3 g of sodium hydroxide. Mix the solutions. Add 10 g of Stractan (Stractan 10 is obtained from Stein-Hall and Company, Inc., 385 Madison Ave., New York), and dilute to 1 liter in a volumetric flask. This solution should be freshly prepared every 3 to 5 days for continuous use.

Procedure

Add 10 ml of the adsorption mixture together with 5 ml of 95% ethanol to a midget impinger and set a sample air flow for a measured rate of approximately 1 liter/min for from 30 min to as long as 2 hr.

When collection is complete, add 1.5 ml of the acid amine test solution and one drop of ferric chloride solution to the contents of the impinger, mixing thoroughly at each addition. Transfer the solution to a 25-ml volumetric flask and add one to five drops of ammonium phosphate solution to complex excess ferric ion. Dilute to volume with distilled water, allow to stand for 30 min, and measure the absorbance at 670 mμ versus a reagent blank of unexposed solution.

Calibration

To a series of 25-ml volumetric flasks, add 10 ml of the absorbing reagent followed by 1, 1, 2, 3, 4, and 5 μg-eq of pure sulfide, which may be either purchased as standard oxygen-free reagent or prepared according to the method of Adams et al.[85] Add 1.5 ml of acid amine test solution. After mixing, add one drop of ferric chloride solution. Mix thoroughly, dilute to volume and allow to stand for 30 min before determining the absorbance at 670 mμ versus a reagent blank.

A standard curve can then be prepared of absorbance versus number of micrograms H_2S per milliliter.

If an H_2S permeation tube is available, a dynamic calibration can be made according to the procedure described in Chapter 5.

Calculation

Using the measured flow in liters per minute and the sampling time in minutes, determine the volume sampled in liters and adjust to 760 mm Hg pressure and 25°C to obtain the standard volume of V_s sampled. Then, using V_s,

$$H_2S, \mu g/m^3 = \frac{(\mu g\ H_2S\ \text{from standard curve}) \times 10^3}{V_s}$$

Interferences

Among the few interferences to this method, strong reducing agents prevent color development, while NO_2 at higher than 0.5 μg/ml produces a pale yellow color that is a positive interference. Sulfides other than H_2S interfere unless Stractan 10 is present in the absorbing solution.

NITROGEN DIOXIDE (NO_2)

Sources of NO_2

Comparison of world-wide sources of nitrogen oxides (NO_x) shows that, of the two most important NO_x pollutants [nitric oxide (NO) and nitrogen dioxide (NO_2)], NO is found in greater quantity. Although most NO is produced by natural bacterial action, it must be emphasized that the rate of emission from man-related sources has, by estimate, increased by one-fourth between 1966 and 1968.[86]

Certainly, man's greatest contribution to the family of nitrogen oxides arises from the combustion of fuels. Although nitrogen is generally thought of as inert, it combines with air oxygen at high flame temperatures to form oxides of nitrogen. Other sources (e.g., chemical and nitric acid manufacture, the detonation of nitrate-containing explosives, and electrical arc processes) contribute very little by comparison.

Relative contributions from various sources are given in Table 3.

Table 3 Summary of Nationwide NO Emissions, 1968[87]

Source	Emissions $\times 10^6$ tons/year	Percent of Total Emissions
Transportation		
Motor vehicles		
(gasoline-powered)	6.6	32.0
Other	1.5	7.3
Fossil fuel combustion	10.0	48.5
General process industry	0.2	1.0
Other man-related sources	2.3	11.2

Air Chemistry of NO_2

Of greatest interest in the air pollution chemistry of the nitrogen oxides are the gas-phase reactions of nitrogen oxides and particularly NO, which reacts rapidly with ozone to form NO_2. This cycle actually begins when some NO_2 present in the atmosphere absorbs sunlight energy to yield nascent oxygen, a very powerful oxidizing agent.

$$NO_2 + UV\ light \rightarrow NO + (O)$$

the reaction then continues to produce ozone

$$O_2 + (O) \rightarrow O_3$$

or intermediate hydrocarbon–oxygen species (see Chapter 2).

Finally, the reaction terminates with the reformation of NO_2.

$$NO + O_3 \rightarrow O_2 + NO_2$$

Since the rates of the above reactions are a function of the actual intensity of sunlight, the reactions and their products assume importance only during those seasons in which zenith sunlight flux is near a maximum, generally between June and September. Also, the close relationship between hydrocarbons, nitrogen oxides, and ozone permits a prediction of daily oxidant values associated with early morning $HC-NO_x$ combinations under conditions of atmospheric stagnation. This limiting relationship is discussed more fully in the section describing photochemical oxidants. It is sufficient to note here that, while excess NO reduces ozone levels by acting as a scavenger for O_3, the resulting NO_2 increases the NO_2/NO ratio, with consequent formation of peroxyacyl nitrates and other smog-associated eye irritants.

In general, observational data show that 0.04 ppm NO_x occurs in coincidence with concentrations of nonmethane hydrocarbons between 0.3 and 1.4 ppm (as carbon). Conversely, 0.3 ppm nonmethane hydrocarbon levels have been shown to correspond to an NO_x range of 0.04 to 0.16 ppm.[88]

Methods for Measuring NO_2

The Griess–Saltzman Colorimetric Method

The most versatile method for measuring NO_2 is by Griess–Saltzman colorimetry. This chemical method is based on the reaction of NO_2 with sulfanilic acid to form the diazonium compound. This compound reacts with N-(1-naphthyl)ethylenediamiamine dihydrochloride to form a reddish pink azo dye whose absorbance at 550 mμ, on 15-min standing, can be used as a measure of NO_2 in the total air sample.

Reagents (ACS analytical grade)

  i N-(1-Naphthyl)ethylenediamine Dihydrochloride Stock Solution 0.1%. One-tenth gram of the reagent is dissolved in 100 ml of distilled nitritefree water. (Nitritefree water is prepared by redistillation after the addition of one crystal each of potassium permanganate and barium hydroxide.)

  ii Absorbing Reagent Into 140 ml of glacial acetic acid, add 5 g of anhydrous sulfanilic acid. Heat slightly to dissolve, cool, add 20 ml of 0.1% stock solution of N-(1-naphthyl)ethylenediamine dihydrochloride and dilute to 1 liter. This reagent will remain stable for several months if kept in a refrigerator in an amber bottle.

  iii Standard Sodium Nitrite Solution (0.203 g/liter). Prepare this standard just before calibration by successive dilutions from a concentrated refrigerated stock solution containing 2.03 g of reagent per liter.

  iv Nitritefree Water If distilled water on hand produces a slight pink color with the addition of a few drops of absorbing reagent, redistill the water using all-glass equipment after adding one crystal each of barium hydroxide and potassium permanganate.

Absorbing Apparatus

The air sample is absorbed in an all-glass fritted bubbler whose frit-size dimensional requirements must be closely adhered to in order to yield results of acceptable accuracy. The apparatus shown in Figure 4 yields an absorbing efficiency of over 95% at a flow rate of 0.4 liter/min, where the maximum frit pore diameter is 60 μ.

Nitrogen Dioxide (NO$_2$)   123

Sampling

Place 10.0 ml of absorbing reagent into a fritted glass bubbler as shown in Figure 4. Using a rotameter, set sample flow at no more than 0.4 liter/min for a time sufficient to develop visible pink color (10 to 30 min). Record the ambient temperature and pressure for later sample volume correction. Allow 15-min standing at 20 to 25°C and measure the absorbance versus a reagent blank at 550 mμ. If an absorbance lower than 1.0 is obtained or if the exposed

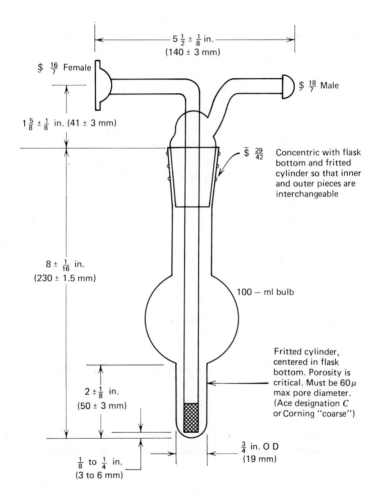

Figure 4. Fritted bubbler for collecting nitrogen dioxide by the Griess-Saltzman method. (From *Methods of Air Sampling and Analysis*, American Publications Health Service, 1972, p. 330).

solution is visibly dark in color, the contents of the bubbler can be diluted to 25 ml. A 1:5 or 1:10 dilution can then be made by diluting a 5.0-ml aliquot of the diluted contents to 25 or 50 ml using absorbent reagent as a diluent.

Calibration

The Griess–Saltzman method can be calibrated using either (1) a solution calibration in which graduated amounts of the sodium nitrite reagent are added to absorbing solution to form a color equivalent to NO_2, or (2) a dynamic calibration in which known air dilutions of NO_2 are made in a stainless steel tank, or a permeation tube (Chapter 5) is used to provide a known flow of NO_2.

Probably the more convenient of these two is the chemical calibration method described here. General directions for dynamic calibration can be found in Chapter 6.

Chemical Calibration Procedure

Into a series of 25-ml volumetric flasks, pipette graduated amounts (0.1, 0.2, 0.4, 0.5, up to 1.0 ml) of sodium nitrite solution and dilute to volume with absorbing reagent. Allow 15 min for development of color and measure absorbance versus unexposed absorbing solution at 550 mμ.

Plot absorbances versus microliters of NO_2 using the relationship

$$10 \ \mu l \text{ of } NO_2 = 10 \text{ ppm } NO_2/\text{liter of sample air} = 1 \text{ ml of standard}$$
$$(\text{at } 760 \text{ mm Hg, } 25°C) \quad\quad NaNO_2 \text{ solution}$$

Join the points so obtained by a straight line through the best fit of points and find the slope K where,

$$K = \frac{\mu l \ NO_2}{\text{absorbance unit}}$$

When using 1.0-cm cells, this value should be approximately 0.73.

Calculation

The concentration of NO_2 in sample air can then be determined from

$$NO_2, \text{ ppm} = \frac{AK}{V}$$

where A is the measured absorbance and K is the slope. Also V is the sample volume in liters per milliliter of absorbing reagent. For example, for a 10-ml bubbler sample,

$$V = \frac{\text{sample volume}}{10}$$

Ordinarily, the absorbance colors are measured within an hour after the required 15-min standing time. For prolonged periods of standing, the absorbance color can be expected to fade by 3 to 4%/day.

The above method is presented as a tentative method of analysis for NO_2 content of the atmosphere by the Intersociety Committee on Method for Ambient Air Sampling and Analysis, No. 43602-01-68T.

Interferences

Negligible interferences have been observed due to ordinary ambient levels of NH_3 or PAN. A fivefold ratio of ozone to nitrogen dioxide is required before only slight positive interference from ozone is observed. Similarly, a tenfold excess of ambient SO_2 has been found to produce no effect on the results, although a thirtyfold excess of SO_3 has been found to cause slight negative interference.

Although interference is normally observed from NO, some NO may be oxidized to actual NO_2 during prolonged sampling time. Therefore, it is desirable to limit the determination of NO_2 to no longer than 1 hr of total batch sampling to ensure retention of NO_2 in the reduced state.

Generally, a precision of 1% of the mean value is observed where limiting factors include air sampling errors and the colorimetry of absorbance measurements.

The 24-hr Field Bubbler Colorimetric Method[89] (Arsenite Method)

Since the Griess–Saltzman method for NO_2 cannot be employed for field sampling or when integrated time sampling of longer than 1 hr is required, a stable field sampling method is necessary. In such situations, it is preferable to use the modified Jacobs-Hochheiser or arsenite method. This method permits

1. Delay in colorimetric analysis for up to 2 weeks after sampling.
2. Sampling periods of as long as 24 hr.

Because the National Air Sampling Network of the Air Pollution Control Office, E.P.A., requires a 24-hr sampling period (on 31 random days per year), the arsenite method has been suggested by the E.P.A. [*Federal Register,* **38** (110), 15,175 (1973)] as a reference method for NO_2. An earlier method, the Jacobs-Hochheiser 24-hr bubbler, allows 1:1 stoichiometry of NO_2 to nitrite ion. However, the collection efficiency of this method for NO_2 is only 35%.[90] Furthermore, some investigators have established substantial interference from NO.[91]

Probably the chief objection raised to the original J–H method is this rather low and variable collection efficiency even under carefully controlled experimental conditions.[89,92,93] In a study by Christie, Lidzey and Radford,[94] it was found that the collection efficiency could be raised from an average 34 to 95% by adding a small amount of sodium arsenite to the absorbing solution. This modification led to the arsenite method, in which NO_2 is collected by bubbling sample air through a basic sodium arsenite solution. The nitrite ion thus formed is reacted with sulfanilamide and N-(1-naphthyl)ethylenediamine in phosphoric acid to form the highly colored azo dye.

Precision and Accuracy

The normal range of the method is from 5 to 750 $\mu g/m^3$ (0.003 to 0.4 ppm) of NO_2, where a concentration of 0.04 μg NO_2/ml yields an absorbance of 0.02 using 1-cm cells. A relative standard deviation of 5% has been observed at an NO_2 concentration of 40 $\mu g/m^3$.[95]

REAGENTS (ACS ANALYTICAL GRADE OR EQUIVALENT)

i Sulfanilamide Twenty grams of sulfanilamide are dissolved in 700 ml of distilled water. Fifty milliliters of phosphoric acid (85%) are then carefully added with mixing and final dilution to 1 liter. When refrigerated, this solution is stable for about 1 month.

ii Absorbing Reagent Four grams of sodium hydroxide and 1.0 g of sodium arsenite are dissolved in distilled water to form 1 liter of solution.

iii Saltzman Reagent Dissolve 0.5 g of reagent N-(1-naphthyl)ethylenediamine dihydrochloride in water to form 500 ml of solution. When refrigerated, this solution is stable for 1 month.

iv Standard Nitrite Solution Dissolve 0.1500 g of sodium nitrite in distilled water and dilute to 1 liter. This solution contains 1000 μg NO^{2-}/ml. When 97% or more concentrated sodium nitrite is used, a gravimetric factor F must be used to obtain the stoichiometric ratio between nitrite ion and NO_2. Here,

$$F = \frac{1500 \times 100}{\text{percent assay NaNO}_2}$$

(*Note.* It is not necessary to weigh exactly 0.1500 g of $NaNO_2$. However, the exact real weight must be entered in place of 0.1500 g when calculating F.)

v Hydrogen Peroxide Dilute 0.2 ml of hydrogen peroxide (30%) to 250 ml using distilled water.

Arsenite Method Absorber System

The absorber cells are usually polypropylene tubes, as shown in Figure 5. Each tube is calibrated to contain 50 ml of absorbing reagent by pipetting a 50-ml aliquot of distilled water into the tube, which is then scribed at the liquid level. Both absorber tubes and fittings must be polyethylene, since rubber stoppers contribute to high blank values. A sample delivery tube that reaches to the bottom of the bubbler is purchased or prepared from 5-mm glass tubing approximately 152 mm in length, with one end drawn out to a jet of 0.6 ± 0.2

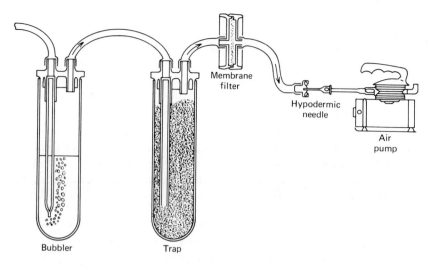

Figure 5. Sampling train, NO_2. (From reference 95.)

mm ID. Sample flow is controlled by a 27-gauge hypodermic needle, which is manufactured to permit a sample airflow of 190 to 210 cm³/min using a pump capable of maintaining a 0.6 atm pressure across the hypodermic needle orifice. Because of the need to maintain exact and constant sample air flow, the needle is calibrated by attaching it as the limiting orifice through which air passes to a calibrated rotameter. The needle must be discarded and replaced with a freshly calibrated needle after the collection of ten 24-hr samples.[96]

Sampling Procedure

The sampling system is assembled according to the arrangement shown in Figure 5.

Fifty milliliters of absorbing reagent are placed in the absorber tube. The flow is tested with the system ready for sampling by installing a calibrated rotameter in place of the funnel at the end of the sample probe. The flow at this time must be at least 85% of the flow rates of needle calibration. After replacing the funnel, restart the pump and sample for 24 hr (EPA suggests midnight to midnight), checking the flow at the end of sample period to determine the average flow.

Colorimetric Analysis

After adding water to make up for evaporation loss during sampling, pipette 10 ml of the exposed absorbing solution into a 50-ml Erlenmeyer flask. Add 1.0 ml of hydrogen peroxide solution to oxidize any absorbed SO_2 to sulfate. Then add, with thorough mixing, 10.0 ml of sulfanilamide solution and 1.4 ml of Saltzman reagent. Treat a 10-ml portion of absorbing reagent in the same way for use as a blank. Allow 10 min for optimum color development and measure the absorbance at 540 mμ versus the reagent blank. Using a standard curve (Calibration Section), read the number of micrograms NO_2^- per milliliter.

Calibration

As described in the Saltzman method for NO_2, either chemical solution or dynamic calibration can be employed. Ordinarily, solution calibration is preferable for field sampling. Here, 5.0 ml of the standard nitrite solution is diluted to 0.2 liter, using absorbing reagent, to form a working nitrite standard of approximately 25 μg NO_2^-/ml (depending on the assay factor F and the actual weight of $NaNO_2$).

Into a sequence of 250-ml volumetric flasks, pipette 1, 2, 5, and 15 ml of the working standard nitrite solution and dilute to volume with absorbing reagent. These solutions, which then contain 0.50, 1.00, 1.25, and 1.50 g NO_2^-/ml. respectively, are placed in 1-cm cuvettes and absorbance is measured versus a reagent blank at 450 mμ. A plot is then prepared in which the x-axis is the number of micrograms NO_2^- per milliliter of absorbing reagent and the y-axis is the measured absorbance.

Calculations

The volume (V) of sample air in milliliters is calculated by multiplying the average sample flow rate in milliliter per minute by the time in minutes. The milliliter volume is multiplied by 10^{-6} to obtain sample air volume in cubic meters.

The ambient level of NO_2 is determined from

$$\mu g\ NO_2^-/m^3 = \frac{\mu g\ NO_2^-/ml\ (measured) \times 50}{0.85\ V}$$

where 50 = volume of absorbing reagent in milliliters
0.85 = empirical collection efficiency
V = air sample volume in cubic milliliters

Parts per million NO_2 by volume can be calculated from

$$\frac{\mu g\ NO_2}{m^3} \times 5.32 \times 10^{-4} = ppm\ NO_2\ (vol)$$

Methods for Measuring Nitric Oxide (NO) and Total Nitrogen Oxides (NO_x)

Since no direct chemical method exists at present for the determination of NO, NO is usually measured by oxidation to NO_2 with subsequent measure of NO_2

formed. Typical oxidizing agents are aqueous potassium permanganate or sodium dichromate. The most frequently used oxidizer probably is a scrubber that is prepared by precipitating chromium trioxide on glass-fiber filter paper.

Preparation of an Oxidizer for Conversion of NO to NO_2

A standard straight or U-shaped drying tube is filled with ¼-in. strips of an ordinary 8 × 10 in. high volume glass-fiber filter media. These strips are prepared and treated in the following manner:

Using scissors or paper cutter, cut in half 10 rectangular high-volume filter papers, and soak for 10 min in a solution prepared by either dissolving 5 g of sodium dichromate and 5 g of concentrated sulfuric acid in 200 ml of distilled water or dissolving 125 g of chromic oxide and 35 ml of sulfuric acid in 750 ml of distilled water.

Remove the filters from the solution and dry in an oven or on a hot plate at 200°F. The filters so prepared should have a slightly pinkish brown appearance.

The active life of such an oxidizer scrubber is limited from a few days to a week to as long as 1 month in atmospheres of low relative humidity (< 20%). A greenish color as well as the appearance of moisture on the filter-oxidizer indicates a "spent" condition. At flow rates between 200 and 300 ml/min of sample air, this scrubber has a reported collection efficiency of 95 to 100%,[97] compared with as low as 70% for the 1 to 5% permanganate fritted bubbler solution. The combination of NO plus NO_2 is usually referred to as "total nitrogen oxides" or NO_x. Because the oxidizer acts to remove any SO_2 that would otherwise interfere 1:1 (i.e., 1 ppm SO_2 removes 1 ppm NO_2) with the color-forming reaction, such scrubbers are commonly used with the continuous or intermittent Saltzman method and only occasionally with the arsenite method, since the latter involves an oxidation step that accomplishes the oxidative removal of SO_2 with peroxide.

Determination of NO and NO_x

It is noteworthy here that the adaptability of these scrubbers permits either intermittent or continuous determination of NO when either sequential or side-by-side oxidizing scrubber and scrubberless systems provide NO_2 and NO_x data, which are then subtracted $NO_x - NO_2 = NO$ to yield hourly average or point-by-point levels of ambient NO.

Oxidants: Ozone and Peroxide

The term oxidants was originally used to define the color-forming reaction equivalent of an air sample with neutral buffered potassium iodide. It is known that a number of organic and inorganic compounds and elements react under the specified conditions to liberate stoichiometric and nonstoichiometric amounts of iodine. Ozone, hydrogen peroxide, chlorine, organic peroxide, and even 10 to 20% of the ambient NO_2 give a positive reaction. For the sake of convenience, this total response has been traditionally recorded as oxidant but reported as ozone.

More recently, because of health studies that attempted to relate specific biological symptoms to ozone exposure, analytical methods have perforce become quite specific for ozone. Since April 30, 1971,[98] with the advent of chemiluminescence methods specific for ozone (Chapter 4), ambient air quality standards have been adopted for ozone, independent of other oxidants. This action is certainly appropriate, since the major atmospheric oxidant is ozone.

Ozone and Its Mechanisms of Formation in Air

Ozone is a product of a rather high-energy oxidation. As such, in its higher energy state, it is capable of causing oxidation in other chemical species. Combustion is also an oxidation reaction whose by-products include suboxides, such as aldehydes, NO, SO_2, and CO. When these contaminants are released into the atmosphere during stagnant weather conditions that promote further interaction, sunlight-induced oxidation reactions continue to produce a variety of mixed oxidation products. Thus during the hours of overhead sunlight (i.e., between 10:00 A.M. and 2:30 P.M.), oxidation reactions convert NO_2 to NO_3^- nitrate, SO_2 to SO_4^{2-} sulfate, and organic aldehydes to peroxyacyl nitrate (PAN), peroxybenzoyl nitrate, and a number of stable organic hydroperoxides.

Thus to make a true assessment of the ambient concentration, we must be prepared to measure both ozone and its atmospheric reaction products. This information eventually will lead us to the conclusion that ozone, which is formed by a number of natural processes, is best controlled by removing those intermediates that would otherwise lead to its stabilization and further buildup in the atmosphere.

Absorption of ultraviolet (UV) light, which does produce ozone in the stratosphere, is not sufficient to explain the presence of ozone at lower levels where man lives, since other gases are even more efficient UV absorbers at these levels. A comparison of various absorption efficiencies of the spectrum of atmospheric components shows that nitrogen dioxide is by far the most efficient absorber. In a complex series of reactions called the NO_2 photolytic cycle, it is possible to understand the phenomenon of ozone stabilization in the lower atmosphere. Here, NO_2 absorbs UV light and decomposes.

$$NO_2 \xrightarrow{uv} NO + O$$

Nascent oxygen then reacts with natural normal oxygen in the presence of unsaturated hydrocarbons to form hydroperoxide intermediates.

$$O + HC{=\!\!=} + O_2 \rightarrow \underset{|}{\overset{OH}{HC}}{-}O{-}O^- \tag{1}$$

with H above the first HC.

These compounds eventually break down to yield free ozone,

$$H{-}\underset{|}{\overset{OH}{C}}{-}O{-}O^- \rightarrow H_2C = +O_3 \tag{2}$$

which reacts with NO to complete the cycle.

$$O_3 + NO \rightarrow NO_2 + O_2 \tag{3}$$

Such series of reactions depend on the presence of a number of intermediates (R—C=) without which completion of the cycle would not be possible. The limiting components could be either NO or hydrocarbons. Leighton[99] suggests that reaction 3 is the rate-determining reaction. Others maintain that this reaction is not sufficient to explain the observed ozone buildup, since both NO and ozone would be consumed in equal quantities.[100] It is therefore assumed that at least both of these reactions and very possibly other reactive species tend to reduce the O_3 concentration in air at the expense of a buildup in the oxysubstituted unsaturates, such as PAN.

In this respect, a number of mechanisms are used to classify various hydrocarbons according to their reactivity.[101] In several studies, yields of peroxyacyl nitrate are reported.[102] A list of photoreactive hydrocarbons is given in Table 2, Chapter 2. In addition, yields of peroxyacyl nitrate, as produced by several researchers, can be found in "Ambient Air Quality Criteria."[103] Photochemical systems such as the above can be produced artificially in the laboratory to provide data relating concentrations of oxidant to yields of various nitrogen and nitrogen carbon derivatives. It is because of the complexity of these reactions and the multitude of products that air monitoring studies have preferred to direct their attention towards ozone and only one of its ambient air derivatives, namely peroxyacyl nitrate (PAN).

Methods for Measuring Ozone

Average concentrations of ozone at ground level seldom exceed 0.08 ppm in average urban air (excluding Los Angeles). Therefore, sufficient sample must be obtained to determine ozone by redox of colorimetry or coulometry.

The most widely used method for oxidant (ozone) determination is probably that of Saltzman.[104] In this method, oxidants are absorbed in neutral, buffered,

1% aqueous iodide solution, with the resulting iodine (as I_3^-) measured colorimetrically at 352 mμ. Here, the ozone-iodine equivalent is 1:1.

Manual Method Using Iodine Colorimetry

Continuous instruments are often operated at iodide concentrations of 10 and 20% iodide to take advantage of increased collection efficiency. However, manual colorimetry, in which a sample is collected in the same solution over an hour's time, requires a much lower 1% iodide absorbing solution to minimize the possibility of induced side reactions that tend to occur when oxidizing species remain in contact with concentrated iodide solutions for any prolonged time.

This technique can be made specific for ozone, as described in requirements for testing ozone in the *Federal Register*,[105] by using a chromic trioxide scrubber placed upstream of the absorbing solution to eliminate negative interferences from SO_2 through conversion of SO_2 to sulfate. Since this scrubber system also converts a portion of NO to NO_2, an additional correction factor of $10NO_x$ must be applied by subtraction from the final oxidant measurement. This, of course requires simultaneous and, if possible, side-by-side measurement of NO_x using a scrubber identical to that employed in the oxide measurement. Certain drawbacks, such as the effects of moisture on the life of the scrubber, have minimized the effectiveness of this method. However, the general application of this approach as a field method for total oxidant is still recommended.

Principle

This colorimetric method involves the liberation of 1 eq of iodine per mole of absorbed oxidant according to the reaction

$$O_3 + 3KI + H_2O \rightarrow KI_3 + 2KOH + O_2$$

This method can be applied to the determination of ambient ozone within the range of 0.01 to 10.0 ppm. Up to 100-fold SO_2, which would otherwise interfere with the determination, can be removed by installing a chromic acid glass-fiber filter (described in the section Measurement of Nitric Oxide). Here, it is well to note that use of the chromic oxide scrubber can be avoided when SO_2 is less than 10% of the NO concentration. This condition will usually prevail unless the sampler is located close to a known SO_2 source, such as a coal-burning power plant. A rule of thumb is that the scrubber can be removed from the system when the SO_2 concentration is below 0.02 ppm, and a direct readout of O_3 will follow without simultaneous determination of either NO_2 or NO_x. Other nonozone but oxidant interferences include peroxyacetyl nitrate, which gives a response up to 50% per molar equivalent of ozone. However, because of the high dependence of peroxide response on retention time in solution, it is difficult to estimate the degree of organic peroxide interference. This interference is not of real concern except in areas such as Los Angeles or other sites of known high concentrations (i.e., 0.1 ppm) of photochemical organic oxidant. Other possible positive interferences include the halogen gases and hydrogen peroxide. Hydrogen sulfide, if present, produces a negative interference.[106]

Precision and Accuracy

Precision of the method lies within 5% deviation. The major source of error results from loss of iodine through volatilization if the sampling period is prolonged for more than 1 to 2 hr unless a second-stage impinger is employed.

REAGENTS: (ACS ANALYTICAL GRADE)

i   Absorbing Solution (1% Potassium Iodide 0.1 M Mixed Phosphate Buffer) Dissolve 10.0 g of potassium iodide in 20 ml of distilled water. Add this solution with stirring to the buffer mixture which is prepared by dissolving, in order, 13.6 g potassium dihydrogen phosphate and 14.2 g disodium hydrogen phosphate in 500 ml water.

Following addition of the potassium iodide solution, dilute to 1 liter, and allow to stand for at least one day. Then measure the pH, preferably with a pH meter or, if none is available, with accurate, fresh pH paper. Adjust the pH to 6.8 ± 0.2 using the sodium hydroxide or potassium dihydrogen phosphate.

ii   Iodine Stock Solution (0.05 N)   To a 500-ml volumetric flask, add 16 g of potassium iodide and 3.173 g of resublimed iodine crystal and dilute to ½ liter volume. Shake and swirl until all but a few crystals of iodine have dissolved, then dilute to volume. Store at room temperature for at least 1 day. Before use, standardize against 0.025 M sodium thiosulfate solution that has been standardized against primary standard potassium dichromate. (This step can be avoided by using commercially available 0.05 N or a 1:1 dilution of 0.10 N iodine solution.)

iii   Iodine Working Solution (0.001 M)   Transfer exactly 5.00 ml of 0.05 N I_2 stock solution to a 100-ml volumetric flask and dilute to volume with absorbing solution. Protect from direct sunlight. Solution is stable for 1 to 2 days.

iv   Calibration Solution   For purposes of calibration, an equivalence can be based upon the relationship where

$$\frac{4.09 \text{ ml of } 0.05 \ N \ I_2 \text{ solution}}{100 \text{ ml}} = \frac{1 \ \mu\text{l of } O_3}{\text{ml}}$$

Dilute (X) milliliters of the measured normality (Y) stock I_2 solution (so that $X - Y = 0.2045$ meq) to 100 ml.

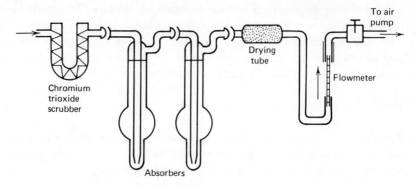

Figure 6. Sampling train for determination of oxidants by the 1% neutral buffered potassium iodide bubbler method. (From reference 106.)

Example. If measured I_2 normality is 0.045, then,

$$X = \frac{0.2045}{0.045} = 4.44 \text{ ml}$$

v SO₂ Absorber   Prepare chromic oxide scrubbers as directed in the section Methods for Measuring Nitric Oxide and Total Nitrogen Oxides. When paper becomes visibly moist, redry in an oven at 80 to 90°C to regenerate.

PROCEDURE. Assemble a sampling system, using either one or two midget impingers, according to the diagram in Figure 6.

The U-tube is filled with a filter consisting of crimped glass fiber impregnated with chromic oxide. Connections should be ground glass or glass joined with butt-to-butt connections using polyvinyl tubing.

Pipette 10 ml of absorbing solution into each midget impinger and adjust sample air flow to 0.5 to 3 liters/min (preferably 2 liters/m) for 15 min. Determine the total volume of air sample and measure the ambient temperature and pressure for use in correcting the sample volume to standard conditions of the method.

Transfer the exposed absorbing solution with minimum washings to a 10-ml volumetric flask. If required, dilute to volume with double distilled water. Within ½ to 1 hr after sampling, measure the sample absorbance at 352 mμ versus unexposed absorbing reagent.

CALIBRATION. Using the calibration solution prepared from I_2 stock solution, pipette 0.5, 1.0, 2.0, and 5.0 ml into successive 25-ml volumetric flasks and dilute to volume with absorbing reagent.

If difficulty is encountered in preparing the calibration solution because of lack of decimal or lambda pipettes, the following calibration procedure can be substituted. Transfer 1.0 ml of the stock 0.05 N iodine solution to a 100-ml volumetric flask and dilute to volume with absorbing reagent. Transfer 1, 2, 3, and 4 ml aliquots of this solution to 25-ml volumetric flasks and dilute to volume with absorbing reagent. The concentration of each solution can be calculated from the equation

$$\mu g\ O_3 = (\text{measured NO} \times 96 \times V)$$

where V = volume of the final aliquot diluted (i.e., 1, 2, 3, etc.)

Read the absorbances at 352 mμ versus absorbing solution and prepare a plot of, micrograms, or microliters of O_3 per 10 ml of absorbing reagent versus absorbance. Draw a straight line connecting the points to produce the best fit.

CALCULATIONS. Correct the sample air volume to 760 mm Hg and 25°C.

$$O_3,\ \text{ppm} = \frac{\mu l\ O_3/10\ \text{ml from calibration curve}}{\text{volume of sample air, liters}}$$

or

$$O_3\ \mu g/m^3 = \frac{\mu g\ O_3\ \text{from calibration curve}}{\text{volume of sample air, liters}}$$

to resolve either units,

$$O_3,\ \text{ppm} = \frac{\mu g\ O_3 \times 5.094 \times 10^{-4}}{m^3}$$

The above method is based on the Intersociety Method for Oxidant (Ozone).[107]

Alkaline Potassium Iodide Method[108]

This field sampling method is subject to the same interferences and equipment requirements as the neutral buffered iodide method and is designed to measure oxidant in the range of 0.01 to 20 ppm.

The main advantage of this approach over the neutral buffered iodide method is that a time delay of several days to 1 week is possible before color development need be produced by acidification of the absorbing solution to yield iodine.

Phenolphthalin Method[109]

When phenolphthalin is oxidized in the presence of copper sulfate, the characteristic pink dye color of phenolphthalin is produced. This relatively rapid method can be calibrated against standard prepared from hydrogen peroxide (H_2O_2-$2O_3$), and suffers no interference from NO_2. Hydrogen sulfide and SO_2 interfere at average ambient concentrations.

Under conditions where no SO_2 or H_2S are present, the method can be used with confidence.

REAGENTS (ACS REAGENT GRADE OR EQUIVALENT)

i Phenolphthalin Solution   Arrange a small reflux condenser by sleeving an air condenser to a 125-ml narrow-necked Erlenmeyer flask. Support the assembly by a clamp around the air condenser and use a hot plate as heat source to reflux for 2 hr a mixture prepared from 10 g sodium hydroxide, 5 g powdered zinc, 1 g phenolphthalin, and 20 ml distilled water. After filtration through a medium-frit sintered-glass crucible, the solution is diluted to 50 ml and stored in an amber bottle containing a few pieces of mossy zinc.

ii Copper Sulfate Solution 0.01 M Prepare by dissolving 0.249 g of copper sulfate pentahydrate or 0.159 g of the anhydrous reagent in distilled water to make 100 ml total volume.

iii Absorbing Solution   To a 100-ml volumetric flask, add 1.0 ml of 0.01 M copper sulfate solution and 1.0 ml of a solution prepared by diluting 10 ml of phenolphthalin solution (BS 1.1) with 30 ml of distilled water. Dilute to volume.

PROCEDURE.   Using one or two midget impingers, fill one (or each) with 10 ml of the absorbing solution. Adjust sample flow to 0.5 to 1.0 liters/min for a period of up to 1 hr. Longer sampling may be required if no color development is observed. (Lack of sample color may also be due to the presence of SO_2.) Measure color in a sample colorimeter using a green filter or use Nessler or comparometer tubes.

CALIBRATION.   Standards for colorimetry or comparometry can be prepared from hydrogen peroxide solutions containing between 0.01 μg and 0.1 μg peroxide/ml. Each standard should contain 0.1 ml each of copper sulfate and phenolphthalein solutions per 10-ml volume. At least 15 min should be allowed for color development before reading in a comparometer or colorimeter.

Gas-Phase Titration Using trans-2-Butene[110]

Selective reaction of *trans*-2-butene with ozone separates this major portion of total oxidant. Thus any iodide methodology including a scrubbing chamber containing an excess of this olefin is capable of providing total oxidants with the scrubber detached and nonozone oxidants through the color formed when the scrubber is in line. Specific ozone concentration follows from the difference between these values.

Rubber Cracking Method for Determining Ozone

Before the use of the iodine bubbler, the standard field method applied by the National Air Sampling Network (NASN) for oxidant depended on the observed ability of ozone to produce, in a flexed rubber strip, cracks whose depth was proportional to the concentration of ozone in the vicinity of the strip. This method was standardized by flexing a rubber strip in the direction of flow of a known part per million oxidant concentration from an ozone generator. Here, anomalies in the results have led the author, as well as other investigators who have employed this method of measurement, to conclude that the method suffers from a number of interferences, such as NO_2 and free radicals of various kinds.[111]

PEROXYACETYL NITRATE (PAN)

A number of studies have been reported that relate the concentrations of NO_2, hydrocarbons, and ozone to the concentration of PAN.[112,113] In studies of the reaction products found on irradiation of saturated, unsaturated, and aromatic compounds in the presence of nitrogen oxides, Altshuller et al.[114] have detected 0.05 to 0.10 ppm of PAN.

Occurrence of PAN

Classically, PAN is formed from reaction of oxidized hydrocarbons in the form of peroxy carbonyls R—(CO)(OO) with NO_2, where

$$\underset{RC-OO^+}{\overset{O}{\|}} + :NO_2 \rightarrow \underset{RC-OONO_2}{\overset{O}{\|}}$$

As reported in the discussion of hydrocarbons (Chapter 2), these compounds are strong eye irritants. As such, their presence in the atmosphere is detrimental to our sense of well-being. Control of these compounds results indirectly by limiting the emission of hydrocarbons and nitrogen oxides from man-related sources, such as hydrocarbons from fuel-tank evaporation, petroleum refining, and paint spray operations and nitrogen oxides from high-temperature combustion sources (e.g., automobiles and power plants).

PAN concentration is a cofunction of oxidant level as a product of hydrocarbon emission and the diurnal variation of sunlight flux. Figure 7 shows the variation of PAN with total oxidant as reported from measurements made in the Los Angeles area. Here, PAN is seen to assume a proportion between 25 and 50% of the total oxidant.

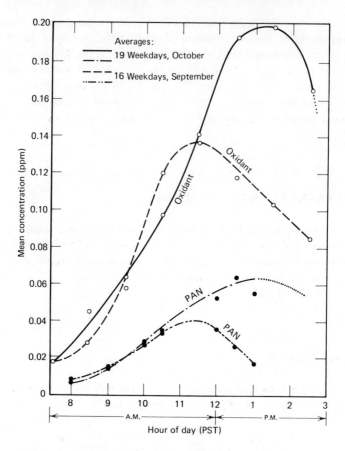

Figure 7. Variation of mean 1-hr average oxidant and PAN concentrations by hour of day in downtown Los Angeles, 1965. (From *Ambient Air Quality Criteria for Photochemical Oxidants,* Section 3, p. 13).

Methods for Measuring PAN

It may be possible through careful plotting of reaction times to separate, by kinetics of reaction, the slow iodide-reacting peroxides from the faster reacting ozone to achieve a differential colorimetric method for PAN. However, gas chromatography, which has classically been used to separate various mixture of homologs into their components, forms the basis for the most rapid straightforward approach in the determination of peroxyacetyl nitrate. Increased sensitivity is achieved by using an electron capture detector, which also makes the method quite specific for PAN.[115]

Chromatographic Method

No sampling or sample preparation is required in this method, which employs a chromatographic column to separate PAN from other components present in ambient air. Retention time under method conditions is about ½ min to 1½ min.

Range and Accuracy

Sensitivity of 1 ppb or lower can be achieved. The only recorded interferences result indirectly when other compounds of high electron affinity (e.g., aromatic hydrocarbons) overload the detector, thereby reducing its overall efficiency. Overall accuracy is judged to be within 5%.

Procedure

A modified procedure based on the method of Darley et al.[115] is proposed by Smith et al.[116] as the Intersociety Committee tentative method for determination of PAN.

Here an Aerograph Model 681 gas chromatograph equipped with an electron capture detector is used together with a 9-in. Teflon column (⅛-in. thick), which is packed with 5% Carbowax E400 deposited on 100 to 120 mesh HMDS-treated Chromasorb-W at 25°C. Nitrogen carrier gas flow is 40 ml/min, and the retention time of the PAN is 60 sec.

Reagents

Reagents include only the described column, purified nitrogen carrier gas, and synthesized PAN, which can be purchased commercially or prepared in the following manner as described by Stephens, Burleson, and Cardiff.[117]

Calibration and Preparation of Standard PAN[117]

Using a 10-cm infrared gas cell with a Pyrex glass wall, flush the cell with pure oxygen. Then, using a gastight syringe, inject approximately 50 μl of ethyl nitrite vapor obtained from the air volume in the cap of a storage bottle containing ethyl nitrite. A fluorescent lamp of UV light 300 to 400 mμ may then be used to irradiate for ½ to 1 hr, during which additional 50-μl amounts of ethyl nitrite are added to the cell at approximately 15-min intervals. No attempt is made to separate PAN formed from other peroxide species. Selectivity *in situ* is achieved by standardization of the PAN content, using an infrared spectrophotometer, with the constant absorptivity of $(13.9 \times 10^{-4})/(\text{ppm})(M)$ used to establish ppm concentration at the measured absorbance through Beer's law:

$$A = abc$$

where A = measured absorbance log (I_0/I)
a = absorptivity, $(13.9 \times 10^{-4})/(\text{ppm})(M)$
b = 0.1-m cell path
c = PAN concentration, ppm

Solving for c,

$$c \text{ (ppm)} = \frac{A}{0.1} \times 13.9 \times 10^{-4}$$

Once determined, this concentration of PAN can be diluted by continuous dilution methods (Chapter 5) or by using 0.100-ml dilution and gastight microliter transfer syringes to prepare 1 to 1000 or 1 to 10,000 dilution rations.

Other Methods for PAN

The calibration method described in method 1, using infrared spectroscopy, can itself be used to determine PAN. The strong absorption wavelength maxima at 5.75 and 8.62 μ can be used with a 500-m gas cell to detect PAN in ambient air.[118]

THE HALOGENS

Occurrence of Halogens

Alkali halide salts (i.e., sodium and potassium fluorides, chlorides, bromides, and iodides) are generally emitted into ambient air from processes that involve the heating of minerals (limestone kilning) or mineral-containing ores (iron refining). Consequently, some processes (e.g., cement manufacture, fertilizer manufacture through the wet-process phosphoric acid method, aluminum refining, and ceramic and glass works) and the combustion of coal and oil, when these contain entrained encapsulated mineral deposits, produce emissions of halides. The volatile and particulate halide compounds of greatest interest from an air pollution standpoint are the fluorides.

The environmental fate of halogens and the halogen compounds generally is passage into ground sediments through precipitation followed by leaching of the less-soluble salts (iodides or calcium fluoride) by acidic ground waters.

Still, it is often necessary to measure ambient halide, especially chloride and fluoride in the presence of a known source of these contaminants.

Methods of Analysis for Halogens

General methods for sampling, which can be applied especially to the volatile halides, involve merely sampling using midget impingers. Sampling is done at a sample air flow rate of 100 ml/min using either one impinger or two impingers in series, with distilled water as the collecting medium. Specific exceptions include the collection of halogen acid vapor and hydrochloric and hydrofluoric acids in 0.1 N aqueous sodium hydroxide solution. Common halide particulates and the volatile sodium and potassium fluorides are easily collected by impingers using distilled water. The more insoluble calcium and strontium fluorides can be collected using a glass-fiber filter. The filter is either placed ahead of a midget impinger, which would then collect the volatile halide to pass through the filter, or be used as a standard high-volume sampler with a flow of 40 to 55 ft^3/min for collection of particulate halide only. Optimum sampling time for the impinger is ½ to 3 hr, while the high-volume sample might be operated for 1 to 6 hr to collect a sample of sufficient size for analysis.

The glass-fiber paper must be Soxhlet or manually extracted with warm water for ½ hr before transferring the solution to a beaker for chemical analysis. An impinger sample can be analyzed directly chemically.

Halogen acid fumes, such as HCl, are known pollutants from pickling baths or acid wash treatment processes. For such fumes, general practice is to collect a 1-hr air sample using a standard impinger at 1 cfm or a midget impinger at 0.1 cfm. The collected sample is subsequently titrated using standard 0.1 N NaOH, with report of the equivalence of sodium hydroxide as total acidity or total HCl or HF.

Other methods for specific halogens are described in the following sections.

Chloride—General Methods

Chloride can be determined titrimetrically by the Mohr or Volhard methods. In the Mohr method, chloride ion is titrated directly with standard 0.1 M silver nitrate solution. Chromate ion, which precipitates to form a reddish particulate with excess silver nitrate at the end point, is used as an indicator. Volhard's method, on the other hand, precipitates the chloride ion with an excess of 0.1 M silver nitrate solution. The excess silver ion is then back-titrated with potassium thiocyanate, using ferric alum as an end-point indicator.

These methods are presented in detail in any standard quantitative analysis textbook. They are mentioned here for the sake of completeness and to refresh the memory of the reader, who, it is assumed, has scientific background.

A method that has been used effectively to measure chloride at low ambient concentrations in the Wayne County Air Pollution Control laboratory employs a modified serum titration using 0.10 N mercuric nitrate solution with an indicator prepared from diphenylcarbazone in ethanol. The titrant is standardized using standard sodium chloride. Chloride ion can then be determined to a few ppm in the air sample using a glass-fiber high-volume filter after a running time of 24 hr at 50-cfm flow rate.

In addition, chloride ion can be determined by o-toluidine colorimetry, ion exchange chromatography, and a specific ion electrode.

Recommended Method for Ambient Chloride[119]

When aqueous dissolved chlorides are titrated with mercuric nitrate solution using diphenyl carbazone–bromophenyl blue indicator, chloride reacts to form the stable $HgCl_2$. At the end point, excess mercuric ion reacts to form a blue-violet complex with diphenyl carbazol–bromophenyl blue.

Range and Accuracy

The procedure is designed to produce a sensitivity of 2 ppm chloride in solution throughout a range extending from 0.01 to 12.0 mg of chloride per 50-ml sample. Interferences include heavy metal ions (e.g., lead and tin) which affect the color of the end point. If the interference is known, such color tones can be compensated for by titration of the "known" containing the appropriate interference at the expected concentration. Precision of 0.1 ppm, or 2% of the higher chloride concentrations, applies to this determination.

Interference from sulfites or sulfides can be removed by the addition of 0.5 ml of 30% hydrogen peroxide per 50 ml of aqueous sample.

Collection of Sample

Gaseous chlorides are collected using 10 ml of distilled water in a midget impinger at a sample airflow of 0.1 cfm for 1 to 3 hr. Particulate chlorides are glass-fiber filtered by high-volume air sampling using a flow of 45 to 55 cfm for 24 hr or are collected by electrostatic precipitation in cases where particulate air loading is especially high (>200 $\mu g/m^3$). The high-volume filter is extracted with warm distilled water.

Sample Preparation of Particulate Chloride Collected on High-Volume Glass-Fiber Filters

Use a 1-in. wide strip of filter (11.1% of the total area) cut into 1×1 in. squares, place in a 50-ml beaker, and add 20 ml of warm distilled water. Allow to stand for ½ hr, mixing occasionally. Then filter through Whatman No. 4 filter paper into a 25-ml volumetric flask and dilute to volume with distilled water.

Reagents

i Mercuric Nitrate Solution (0.014 N) To 25 ml of distilled water containing 0.25 ml of concentrated nitric acid, add 2.42 g of mercuric nitrate and dilute to 1 liter (1 ml = 0.5 mg Cl^-).

Standardize the solution by the titration of a 10.00-ml aliquot of 0.025 M NaCl. Here, the formal mercuric nitrate concentration follows from

$$\frac{N}{[Hg(NO_3)_2]} = \frac{\text{ml NaCl} \times N \text{ (NaCl)}}{\text{ml Hg(NO}_3)_2}$$

ii Standard Sodium Chloride (0.025 N)   Dissolve 1.4613 g of oven-dried (110°C) reagent NaCl in water and dilute to 1 liter.

iii Mixed Indicator   In 100 ml of methanol, dissolve 500 mg of diphenyl carbazone and 500 mg of powdered bromophenyl blue. Transfer to an amber glass bottle and refrigerate. Solution remains fresh for up to 6 months. (This mixture is covered by U.S. Patent 2,784,064.)

iv Sodium Hydroxide Solution   Weight 10 g of sodium hydroxide into a 1-liter flask and dilute to volume with distilled water.

v Hydroquinone Solution   Mix 1 g of hydroquinone in 10 ml of distilled water, stir to dissolve, and dilute to 100 ml.

Procedure

Into a 250-ml Erlenmeyer flask, place an aliquot (less than 50 ml) of aqueous chloride containing a sample from the impinger or filter extract. (*Note.* The aliquot should contain less than 12 mg of Cl^-.) Five drops of the mixed indicator are added, followed by 2 ml of freshly prepared hydroquinone solution, and the color of the solution is checked to determine pH in the following manner. A red to blue-violet solution is too basic. Add dilute nitric acid (1:300) dropwise to change color to yellow, usually about five to eight drops. While mixing, continue to add dropwise 1.0 ml in excess. Here, using narrow range pH paper, check to determine if the pH is 3.0 to 3.5.

Using standard mercuric nitrate solution, titrate dropwise, waiting 30 sec between drops, until a blue-violet end point is observed by transmitted light. Use of a microburet is advisable.

Calculation

When the sample volume in milliliters equals (S), subtract any volume in milliliters (B) required to titrate a distilled water blank so that $S - B = V$ in milliliters. Proceed to calculate:

$$\text{ppm HCl} = \frac{VNmT \times 6.24 \times 10^4}{APm}$$

where N = the normality of the Hg $(NO_3)_2$
M = total volume of sample collected
m = volume of sample aliquot
T = sampling temperature, °K
P = atmospheric pressure at sampling time, atm
A = volume of air sampled, liters.

Fluoride—General Methods

Colorimetrically, fluoride can be determined by its bleaching effect on several organic dyes such as zirconium alizarin and zirconyl-Eriochrome Cyanine R.

As low as 1.5 ppm aqueous fluoride can be determined by its bleaching effect on aluminum-hematoxylin.[120]

Since polyethylene is inert to attack by fluoride ion, sampling equipment for HF, especially, may well be fabricated of polyolefin material to prevent loss of sample due to reaction with the walls of the container.

The most reliable quantitative determination of fluoride probably is contained in a procedure summarized in the presentation of tentative methods for fluoride analysis in Health Laboratory Science,[121] which contains separation and analysis methods.

Recommended Method for Ambient Fluoride

Sampling Methods

Place 10 ml neutral or basic aqueous collecting medium in a midget impinger. After 1 hr at a flow rate of 1 cfm, transfer the aqueous neutral or 0.1 M alkaline collecting medium from a midget impinger to a 50-ml beaker and add two or three drops of phenolphthalein. Make basic to phenolphthalein, if required, using 0.1 M NaOH, and add five drops of 30% hydrogen peroxide. Boil to destroy excess peroxide.

Method of Sample Preparation for Fluoride Collected on High-Volume Glass-Fiber Filters

Fiberglass filters, 45 to 55 cfm, 24-hr sample.

Filters that have been treated with sodium hydroxide to absorb fluoride may be water leached, treated with 30% hydrogen peroxide, and evaporated to 10 ml.

Glass-fiber filters that are untreated should be transferred to a distillation flask for Willard-Winter isolation of fluoride.

General Fluoride Sample Pretreatment

SEPARATION OF FLUORIDE FROM COLLECTING MATRIX. If the sample contains particulate fluoride, which is soluble in concentrated acid (normally any fluoride), it must be separated from the sample matrix, whether dry or wet, by distillation (Willard, Winter distillation).

The prepared matrix or evaporated aqueous solution is dissolved in sulfuric or perchloric acid and steam distilled over soft glass beads to volatilize the fluoride as silicon tetrafluoride into a receiving flask at $135 + 2°C$ until 250 ml of distillate are collected (about 2½ hr) from a sample distilling flask containing about 50 ml of the aqueous matrix. Excess water is a product of the steam-generating flask, which is treated with sodium hydroxide and phenolphthalein to remain alkaline at all times during the distillation.[121]

Recoveries from this separation vary within 10% of the amount of fluoride present with a standard deviation of about 2.5%.

To minimize interferences, large amounts of chloride, if known present, should be removed prior to distillation by precipitation with silver perchlorate solution. In addition, samples containing large amounts of aluminum, boron, or silica usually require a distillation at 165 rather than 135°C.

Although ion exchange methods have been applied to the isolation of fluoride from aqueous impinger or bubbler solutions with some success,[122] distillation remains most generally applicable method of F^- isolation.

Spectrophotometric Analysis of Fluoride

While reaction of fluoride with the lanthanum alizarin lake results in increased absorbance, most of the more widely used reagents, such as zirconium-Eriochrome Cyanine R and zirconium-SPADNS, experience a color fading in proportion to the fluoride concentration.

RANGE AND SENSITIVITY. In this spectrophotometric method, interferences generally are at a minimum because of the prior isolation step of the Willard, Winter distillation or iron exchange separation of fluoride from the sampling medium. The usual detection limit of fluoride is in the range of 0.02 $\mu g/ml$, with straight-line Beer's law response over the range of 0.00 to 1.40 $\mu g\ F^-/ml$.

REAGENTS FOR THE ZIRCONIUM—SPADNS METHOD

    i SPADNS Solution Dissolve 0.985 g of SPADNS [4,5 dihydroxy-3-(p-sulfophenylazo)-2,7-naphthalene disulfonic acid trisodium salt] in distilled water and dilute to 500 ml.

    ii Sodium Fluoride Stock Solution (1 ml = 1.0 mg F^-) Using cp 100% sodium fluoride, dissolve 2.2105 g in distilled water and dilute of 1 liter.

    iii SPADNS Reference Solution To 100 ml of distilled water, transfer 10.0 ml of SPADNS reagent and acidify using 1:1 hydrochloric acid.

    iv Sodium Fluoride Working Solution Transfer 5.0 ml of the sodium fluoride stock solution to a 500-ml volumetric flask and dilute to volume. Store in a polyethylene bottle.

    v Zirconium Reagent Solution To 0.265 g of zirconyl chloride octahydrate ($ZrOCl_2 \cdot 18H_2O$) in 50 ml of water, add 700 ml of concentrated hydrochloric acid and dilute to 1 liter using distilled water.

    vi Zirconium-SPADNS Reagent Combine 50 ml of SPADNS reagent with 50 ml of zirconium reagent. Cool and store in polyethylene bottle up to several months.

PROCEDURE. Dilute an appropriate size aliquot (from 1 to 10 ml depending on expected fluoride concentration) to 25 ml, and add 5.0 ml of the zirconium-SPADNS reagent. After mixing, allow to stand until cooled to room temperature (about 30 min). Transfer to spectrophotometer cell and read absorbance at 570 mμ versus a blank of SPADNS reference solution.

CALIBRATION AND CALCULATIONS. Using a series of 25-ml volumetric flasks, add 5 ml of Zr-SPADNS reagent to each flask and add 0.5, 1.0, 2.0, and 3.0 ml of fluoride working solution in sequence. Allow to stand for about 30 min and measure the absorbance versus a SPADNS reference solution blank at 570 mμ. Plot the absorbances versus μg F^- and determine actual F^- from standard curve using absorbance of sample solution.

Other Methods for Fluoride

Although both a direct and a back-titration procedure are used for samples containing between 0.05 and 10 mg fluoride, the direct titration method is often more easily applied.

In the direct titration method, standard thorium nitrate is used to titrate fluoride at pH 3.0 to an end point at which a slight excess of thorium nitrate forms a pink thorium alizarin lake.[123]

In general, if merely an indication of fluoride is desired, a sample of suspected fluoride material from dustfall, or perhaps a soiled surface can be added to the Zr-SPADNS reagent and visual evidence of bleaching used to confirm or deny the presence of fluoride. It must be kept in mind that any sulfite or similar crystalline reducing agent will tend to bleach the dye, thus causing negative interference.

METALS AND METAL IONS

In the past few years, several metals and metallic compounds have been identified as air contaminants, and regulatory legislation has been enacted by the Federal Government to limit their emission into the atmosphere.

The first three such particulates to receive individual attention were mercury, beryllium, and asbestos, and emission standards for these metals were published on December 7, 1971.[124] In addition to these contaminants, several other metallic inorganic particulates are treated in this section due to obvious concern of many researchers in the field of air pollution chemistry and biology.

Sources and Description of Metallic Particulates

Mercury

It is thought that most mercury compounds are degraded into elemental mercury or oxidized rapidly to mercuric oxide under the action of sunlight. In either case, this airborne particulate can be inhaled by man directly or can be ingested through water or food. It has been suggested that more than 75% of inhaled mercury is absorbed, at concentrations exceeding 50 $\mu g/m^3$.[125] The central nervous system is the area of greatest concern in long-term exposure to mercury at low to moderate levels. Such compounds enter the atmosphere through volatilization from smelters and from the combustion of coal and other fossil fuels.

Another critical source of emission is the process of chlorine manufacture in which a contact cell of metallic mercury is employed as the electrical pole on which chloride is reduced to elemental chlorine. Although waste mercury from this process is considered by some to be relatively inert, it has long been suggested as a possible source of water contamination. Only recently has the mobility of mercury in the environment come to be appreciated through the discovery of the bacterial methylation mechanism, which permits the converting of insoluble mercury into volatile dimethyl mercury. Thus otherwise insoluble mercury can become incorporated in the body tissues of aquatic life as the soluble alkyl compound.

Recent data suggest present ambient air concentrations of ten to several hundred nanograms per cubic meter concentration in ambient air, especially in the vicinity of large cities.

According to National Emission Standards for Hazardous Air Pollutants, the proposed standard of no more than 5.0 lb emission per 24-hr period is an attempt to control only those industries that employ large amounts of mercury in their normal production processes. As explained here,[125,126] peripheral users of mercury would not be expected to contribute significantly to the background ambient level.

Asbestos

Of the several diseases to be associated with inhalation of asbestos particulate by humans, the first was asbestosis, a nonmalignant fibrotic disease. Recently, several researchers, including Selikoff and Wright, have related exposure of asbestos workers to incidence of lung cancer.[127]

One of the most prevalent urban sources of asbestos is the spraying of this material on the surface of steel construction support beams as a heat retardant in case of fire. (Of secondary interest have been particulates resulting from normal wear on automobile brake linings.) Several cities (e.g., New York, Chicago, San Francisco) have enacted ordinances prohibiting such spraying. Coupled with the Federal source standard prohibiting visible emission of asbestos particulate, and the possibility of using substitute materials such as rockwool spray, it is conceivable that a great part of this air contamination will soon vanish. Meanwhile, the city of Chicago, for example, has set a source standard of 2 fibers/cm^3.[128]

Beryllium

As a consequence of an investigation into symptoms experienced by a group of occupationally exposed workers in 1940, two forms of beryllium-related lung diseases were identified. One is an acute form of chemical pneumonitis, and the other is a chronic and progressive disease (berylliosis) involving the alveolar walls.

In 1949, the Atomic Energy Commission developed a rough guideline for beryllium exposure in community air. Other research based on best estimates of airborne beryllium concentrate in the community suggested that levels of metal below 0.01 $\mu g/m^3$ were probably not associated with the beryllium-related diseases.[129,130] Consequently, based on the recommendation of the Committee on Toxicology of the National Academy of Sciences, the Federal Government has established an ambient air quality standard for beryllium of 0.01 $\mu g/m^3$ for 30 days.

In contrast to mercury, whose volatility leads it to migrate great distances from its source, airborne beryllium is quite source related. Therefore, its control is important for known sources, such as grinding, alloying, and cutting operations, that employ this metal because of its hardness and resistance to high temperatures.

In 1966, the inclusion of beryllium in rocket propellants was prohibited. With the cooperation of those industries in the recently enacted federal legislation, it is expected that little, if any, danger should exist from this particular inorganic contaminant in the future.

Lead

The concentration of lead in ambient air has been monitored from the time tetraethyllead was added to gasoline. In most cities in the United States, the annual average concentration ranges from somewhat less than 1 $\mu g/m^3$ to about 4 $\mu g/m^3$. The 1969 NASN quarterly data had a median value of 1 μg Pb/m^3, with under 2 μg Pb/m^3 at 90% of all sites.[131] The concentration in rural areas is about 0.1 μg Pb/m^3, while sampling done in extremely isolated areas gives values as low as 0.0001 $\mu g/m^3$. Emissions from smelters and metal-reclaiming operations provide strong local sources of lead. Nevertheless, the close correlation of lead in air with traffic volume and distance from streets makes it clear that motor vehicles are responsible for most of the lead in air.

The acute toxicity of lead is well known. The problem of acute toxicity seldom occurs today in the industrial situation. The major portion of the problem today comes from young children who eat chips of leaded paint from the walls of older homes. There has recently been a great deal of interest in subclinical effects of lead at moderately elevated blood lead levels.

Hernberg et al.[132] and others have found that ALA-dehydrogenase levels in mature red cells are inversely related to the concentration of lead in the blood. Although ALA-dehydrogenase is an enzyme active in heme synthesis, it is not active in mature red cells. At this time, no adverse health effect has been shown for somewhat lowered ALA-dehydrogenase.

Several investigations have suggested that children eating leaded paint may be additionally exposed due to lead in air and dust.[133] Other authors have reported work showing that lead in air and dust are not important sources.[134] About 10% of ingested lead may be absorbed by the stomach, while 20 to 50% of the lead in air may be absorbed by the lungs. The small particle size of airborne lead aids its penetration of the lung. It is clear, however, that lead in paint is an overwhelming source compared to lead in air.

It seems probable that the use of lead in gasoline in the United States and in some other countries will be reduced because it poisons exhaust catalyst systems used to control automotive emissions.

Cadmium

One of the rarer metals in the earth's crust (0.5 ppm), cadmium is geologically associated with lead and zinc. Thus in the practical order of refining and fabrication, its presence as an air contaminant is largely associated with uses of these two metals.

This element has a very long biological half-life in man, on the order of 10 to 25 years, as compared with as short a time as $2\frac{1}{2}$ months for dimethyl mercury.

Although known exposure to cadmium has been restricted to workers in lead and zinc industries, Japanese farm workers who were exposed to cadmium in runoff water from a zinc mine were found to have developed a bone-embrittling disease known as osteomalacia. Later research showed some evidence of a relationship between cadmium compounds in the human body and incidence of heart disease.

In an effort to determine more exactly the degree of seriousness posed by cadmium in the urban environment, the Environmental Protection Agency has commissioned several studies to determine the extent of cadmium in the environment and its relationship to or endangerment of health.

Miscellaneous Toxic Metals

Although some 27 metals are considered more or less toxic, possible air pollution hazard has thus far been considered only from levels of mercury, nickel, cadmium, lead, beryllium, and antimony, with the need to monitor iron and copper dictated mainly by the relationship of these metals to specialized industrial sources.

The mention of atmospheric nickel results from the toxicity of its volatile carbonyl, which is emitted during lean-air combustion of fuels reformed over nickel oxide catalysts. Antimony, which is highly volatile as well as toxic, is found as a metamorphic metal together with copper and lead and enters the atmospheres as a cocontaminant when these less-volatile, but more common, metals are subject to mechanical wear.

Gravimetric Measurement of Total Suspended Particulate

Filters, which have been desiccated at room temperature for 24 hr (or oven-dried for 1 hr and desiccated), are exposed to a high-volume sample flow of 45 to 55 cfm for 1 to 24 hr. They are then redesiccated for 24 hr (or oven-dried if such was the case in preparation) and reweighed to the nearest 0.1 mg. Sample air volume is determined by multiplying the average flow in cubic feet per minute by time to determine the number of cubic feet of sample air. This volume is then corrected to cubic meters by dividing:

$$V, \text{ m}^3 = \frac{\text{total cubic feet}}{35.3 \text{ ft}^3/\text{m}^3}$$

The normal expression of this air-suspended particulate is micrograms per cubic meter (1 μg = 0.001 mg).

Acidity or Basicity of Supended Particulate

The exposed high-volume glass-fiber filter is extracted with distilled water using either a Soxhlet extractor (½ hr) or a 250-ml beaker into which the folded or divided filter is placed together with 100 to 150 ml of water on a hot plate (85°C) for 1 hr.

The resulting solution is then filtered rapidly through Whatman No. 1 filter paper and titrated with either 0.01 M HCl or 0.01 M NaOH (depending on a spot test of the pH of the solution) to determine acidity or alkalinity. Compensation for the acidity of the filter media glass fiber as well as paper) must be made by duplicate extraction procedure to produce a "blank."

If a greater volume of reagent is used to titrate the "blank," it must be assumed that the acidity or basicity of the sampling and filtering media is greater than that of the air sample obtained, and an expression of this finding is shown in the results.

Ordinarily, acidity as milligrams of HCl follows from

$$\text{acidity (mg HCl)} = \frac{VN36.5}{A} = \text{mg/m}^3$$

where V = (ml titrant) − (ml blank)
N = actual normality of the base used
A = volume of sample air in m³ corrected to 76 mm Hg and 25°C.

Ring Oven Method

The ring oven technique was first introduced as a method for conducting separations on a single piece of filter paper so that subsequent quantitative analysis could be performed on the separate components. Each analytical method is quite specific and can be used as a detection approach that is sensitive to as small as nanogram quantities of inorganic ions. Furthermore, after the spots are separated, they are amenable to microscopic examination and can be identified by crystallography.

This method was first introduced by Weisz in 1954.[135] Initial applications made use of spot test sprays to qualitatively distinguish components in inorganic mixtures.

In light of this application, what greater inorganic mixture can be imagined than that found as ambient aerosol particulate? And, by reason of this, what better application of the ring oven technique? Later, researchers[136] pointed out the usefulness of the method as a quantitative approach. Since then, this relatively inexpensive method of analysis has been used (most often in academic institutions) to demonstrate the ease of inorganic particulate identification and quantitative analysis.

Description of Analysis by the Ring Oven Technique

When a piece of filter paper is placed on the heated 22-mm ID ring of the oven and a drop of solution to be analyzed is placed in the center of the paper, centered on the ring, the sample migrates through the pores of the paper to a point as far as the heated ring, where the solvent (water) is evaporated and the crystalline sample particulate is fixed to the paper.

The filter paper can then be cut into quadrants, each of which can be treated with a reagent specific for the identification of a particular anion (e.g., sulfate, nitrite, iodide or such cations as copper, nickel, or iron).

The reader with a background in chemistry will recall spot tests such as those referred to in Chapter 1 for use under the microscope, in which iron forms the red $Fe(SCN)_6^{3-}$ following treatment with dilute potassium thiocyanate in 0.1 M hydrochloric acid or nickel forms the red nickel dimethylglyoxime. Here, reagent sprays and "crayons" prepared from glycerol stearate containing organic reagents such as dithizone (for lead) and dithiooxamide (for copper) are used to apply a selective reagent to a section of paper.

After identification of an ion is confirmed by spot test methods, quantitation can be achieved by preparing standards for color comparison or by aqueous extraction of the paper followed by standard volumetric or spectrophotometric chemistry.

A number of specific tests can be found in the works of West[137,138] for metals and anions commonly encountered as air pollutants.

Atomic Absorption Methods

Use of the actual spectral lines emitted from a hollow cathode by the element to be determined separates atomic absorption from other methods of element emission spectrometry and permits extremely selective determination of most of the elements on the periodic chart.[139]

In principle, actual insoluble molecular compounds, which satisfy the definition of having a representative population of atoms of the element to be determined, can be admitted into the exciting flame. However, it is usually advisable to dissolve the sample, to permit the addition of chelating and/or ionization suppressants, which enhance the sensitivity of this method of approach. Either vapor discharge or hollow cathode lumps are available as radiation sources. Depending on combinations of spectral lines, as many as five or six elements can be excited from the same lamp.

In the usual mode of operation, spectral radiation produced by the source lamp is selected for the highest-intensity spectral line to be used for photometry. This line can be chosen on the basis of its intensity alone or as a less intense line because of its remoteness from some other element of potential interference in the spectrum.

Sample preparation involves only the extraction of the particulate sample from the collecting filter medium. Once this is accomplished and any ionization suppressant salt added as directed in the manufacturer's instruction manual, no special background is required of a technician, who then merely introduces the sample solution directly from its dilution flask into the flame by means of a capillary tube.

Acetylene–air is usually used as a flame combustion mixture. However, for determinations involving rather refractory metals (e.g., beryllium or titanium), the higher atomizing temperature required is produced by substituting nitrous oxide for air as the oxidizing gas.

Determination of Metals on High-Volume Filter Media Using Atomic Absorption Spectrophotometry

Sample Collection[141]

Desiccated, tared high-volume glass-fiber filters are used to collect airborne particulate samples. Suggested collecting time is 24 hr at a flow rate of 45 to 55 cfm.

Sample Preparation

Strips 2 to 2½ in. wide are cut across the width of the paper, and the exact square inches (or percentage of the total square inches of filter) are used to determine the representative portion of sample weight. Here, using individual filters, approximately 15 to 20 in.^{2} of fiber mat provides enough sample from an average air sampling where the total high-volume glass-fiber filter contains approximately 63 in.^{2} of exposed surface.

Extraction is carried out preferably using a Soxhlet extractor with refluxing for a total of 6 to 8 hr. A 125-ml extraction flask is used or, in the case of composite samples, a 250-ml flask.

When composite samples are used, at least 10 in.^{2} of paper should be included from each filter. As many as six such strips usually can be extracted simultaneously using the same flask.

A few boiling stones are added together with 50 to 75 ml of 1:1 nitric acid. When complete pulping of the filter medium is observed (6 to 8 hr), the mixture is suction-filtered through a medium-porosity sintered-glass crucible. After at least three washings of the filter, it is transferred to a 400-ml beaker and evaporated to incipient dryness.

The sample is allowed to cool and moistened with 2 ml of concentrated nitric acid. After rewarming to dissolve any crystalline material, the sample is then diluted with 20 ml of distilled water and mixed to dissolve any residue.

The solution is then filtered through Whatman No. 42 filter paper, and the filtrate is collected in a 100-ml volumetric flask. If the solution remains cloudy because of pulverized glass fibers, it should be refiltered to remove this particulate, and thus prevent clogging of the capillary of the atomic absorption spectrophotometer.

Once the solution appears clear, dilution is made to 100 ml with distilled water.

General Analysis Procedure

The prepared sample is directly admitted into the flame of the atomic absorption spectrophotometer by means of a capillary outlet.

It is advisable to allow hollow cathode tubes to warm up for at least ½ hr prior to use. This can be done either with the tube in the spectrophotometer or on a multiposition electric rack unit, which is designed to energize several tubes simultaneously. Such a unit is most helpful when several elements are to be determined from a single filter or a single composite of such filters.

The instrumental manual should be consulted to determine the correct wavelength setting for each element. However, example settings are presented in Table 4, together with typical 100-ml solution concentrations and ambient air values for several selected metals.

Table 4 Wavelength Settings and Typical Values for Atomic Absorption Determination of Airborne Metals[140]

Metal	A° Wavelength	Solution Concentration (mg/100 ml)	Metal in Air[a] ($\mu g/m^3$)
Zn	2138	5.50	1.0300
Fe	2967	24.90	4.6600
Ni	2320	0.13	0.0240
Pb	2833	5.50	1.0500
Cd	2288	0.04	0.0073
Be	2348	0.01	0.0020

[a] 1971 annual average.

Calibration and Calculations

Solutions of known metal ion concentrations are used to obtain meter deflections from which a standard curve can be plotted. Milligrams of metal ion per milliliter of filter extract can then be read directly from the curve after appropriate blank correction, if necessary.

Methods of Measurement for Common Specific Metal Ions

Mercury Determination[132]

The method for determining mercury as mercury vapor, mercury oxides, or other compounds usually involves collection of an air sample using either a high-volume glass-fiber filter or a sampling train made up of standard impingers to which is added 100 ml of acidic iodine monochloride solution. The collected mercury is reduced to the metallic state by hydroxylamine sulfate in basic solution. Mercury is then vaporized directly from solution by bubbling with carrier gas (zero Hg air) into the path of a mercury vapor meter or atomic absorption spectrophotometer (flameless operation).

Reagents (ACS Reagent Trade or Equivalent)

ABSORBING SOLUTION (0.1 M IODINE MONOCHLORIDE). To a 2-liter beaker, transfer 800 ml of 25% potassium iodide solution (dissolve 250 g of potassium iodide in 500 ml of distilled water and dilute to 1 liter). To the same beaker, add 800 ml of concentrated hydrochloric acid, mix thoroughly, and allow to cool. With continuous stirring, add very slowly 135 g of potassium iodate.

Continue stirring until free iodine has dissolved and the solution assumes a clear reddish orange color. Dilute this solution to 1800 ml and transfer to an amber bottle for storage.

Withdraw 100 ml of the above solution and dilute to 1 liter to yield working absorbing solution.

REDUCING REAGENT. Dissolve 12 g of hydroxylamine sulfate and 12 g of sodium chloride in 50 to 60 ml of distilled water and dilute to 100 ml.

MERCURY CALIBRATION SOLUTION. Dissolve 0.1354 g of mercuric chloride in 80 ml of 0.3 N hydrochloric acid and dilute to 100 ml with 0.3 N hydrochloric acid (1 ml = 1 mg Hg).

Procedure

Using midget impingers, each containing 15 ml of the absorbing solution, set the flow rate at 0.1 cfm and sample for 1 to 3 hr.

If a high-volume sample is used, the 24-hr sample provides sufficient particulate. However, there is presently some question about the overall collection efficiency of the glass-fiber filter mat for mercury vapor.

All glassware used must thoroughly be cleaned and rinsed with 1:1 nitric acid prior to use.

Using an atomic absorption spectrophotometer whose electronics have been allowed to warm up for at least 30 min, set the absorption wavelength for 253.7 mμ. Bypassing the sample tube, allow the gaseous effluent from a mercury sample tube to flow directly into the burner channel according to procedures in the instrument manual.

Treat 50-ml aliquots of both the sample solution and calibration solution (0.1, 0.4, and 0.6 μg/ml) with 5 ml of 10 N NaOH and 5 ml of reducing reagent, stopped and shaken before placing in the sample line to produce the free mercury vapor. Calculate mercury using a curve prepared from known standards.

Beryllium Determination

Beryllium is determined by standard atomic absorption spectrophotometry after digestion in sulfuric acid. Refer to instrument manufacturer's manual for specific directions.

Selenium Determination[142]

The compound hydrogen selenide (H_2Se), an isostere of H_2S, is emitted into the air from generally the same sources as H_2S (i.e., metal smelting and coking). Because of the very short life of H_2Se in the atmosphere, it is common to find selenium in either the oxide or elemental form.

Samples of airborne selenium, suitable for analysis, can be collected on cellulose or glass-fiber filters, which are then extracted with nitric acid or a mixture of 10 HNO_3 : 1 $HClO_4$ to oxidize selenium compounds to Se(IV). Selenium as Se(IV) reacts with 2,3-diaminonaphthalene in acidic solution to form red fluorescent 4,5-benzopiazselenol, which is then extracted with toluene. The fluorescence is measured at 5.90 mμ with spectroexcitation at 390 mμ.

This fluorometric method permits detection of less than 0.2 $\mu g/m^3$ in ambient air. Other metal interferences can be minimized by passing the sample solution through Dowex 50 W×8 exchange resin.

Antimony Determination[143]

Particulate containing antimony is collected using cellulose or glass-fiber filters. The filter is then placed in a 125-ml flask and treated with nitric–sulfuric acid followed by perchloric acid to oxidize antimony compounds to Sb(V) for subsequent reaction with Rhodamine B. Phosphoric acid is added to suppress iron interference, and benzene or toluene is used to extract the organic phase of the colored complex for measurement in a spectrophotometer at 565 mμ.

Use of isopropyl ether as the extraction medium further reduces interference by iron when this is the only interference of note.

The method is sensitive to 1.0 μg of antimony with a lower detection limit of 0.05 $\mu g/m^3$ in ambient air.

Manganese Determination[144]

The most suitable nonatomic absorption method for determining airborne manganese is based on the treatment of 1 to 2 in. of 24-hr-exposed high-volume fiberglass filter medium with nitric and sulfuric acids. These acids volatilize any chloride present to chlorine and prepare the sample for oxidation with potassium periodate to potassium permanganate, which is measured at 525 mμ in a spectrophotometer.

Using a 2-cm cuvette, 3 $\mu g/m^3$ of manganese can be detected using this method. A longer light path is required for lower concentrations.

Interferences include zinc (above 0.30%) and ferrous salts. The latter can be eliminated by digestion with excess nitric acid.

Molybdenum Determination[145]

Atmospheric concentrations of molybdenum are usually below 0.34 $\mu g/m^3$. This metal is found in atmospheres surrounding steel making and the fabrication of alloy steel.

The wet-chemical determination of this element involves collection of particulate using either cellulose or glass-fiber filter medium. Extraction of the particulate is performed using nitric acid, and the stable molybdenum(V) thiocyanate complex is formed using 25% KSCN solution. The complex is then extracted with methylisobutyl ketone, and the absorbance is measured in a 10-cm cell at 500 mμ.

The method is sensitive to 1.1 μg of molybdenum, with a resulting capability of detecting 0.0555 μg/m^3 Mo when the 10-cm spectrophotometric cell is employed. Smaller cells raise the detection limit by a factor of 10 per cell length in centimeters.

Interferences are observed from the platinum metals as well as selenium and tellurium. However, such interference occurs at well above average atmospheric levels of these elements.

Arsenic Determination[146]

With present spraying of insecticides containing arsenic compounds, it is necessary to have reference to a method for arsenic. This element is found in air as either the oxide or substituted arsine. It can be collected as particulate using a high-volume sampler or cellulose filter. Arsenic compounds are then dissolved at low temperature (below 90°C) in hydrochloric acid and reduced to arsenic(III) using potassium iodide and stannous chloride.

The sample is then transferred to a Gutzeit generator, and arsine is volatilized using zinc in hydrochloric acid. Arsine is then passed through glass wool impregnated with lead acetate to remove H_2S and collected in an absorbing solution of silver diethyldithiocarbonate in pyridine to form the red complex, whose color intensity is measured at 535 mμ.

Sensitivity of this method permits measurement of 0.1 μg/m^3 as a lower detection limit.

Interferences include antimony, which forms a diethyldithiocarbonate having an absorbance maximum at 510 mμ. Copper, nickel, chrome, and cobalt also interfere, and their presence should be confirmed if greater-than-normal ambient concentrations are suspected.

Determination of Iron[147]

Samples of particulate that have been collected by filtration using glass-fiber or cellulose filters can be analyzed by titration of the hydrochloric acid extract of the filter medium, using ceric sulfate as the titrant.

If sampling is in the vicinity of an iron fabrication or melting operation, air-filtered samples can well exceed 40 mg iron. If this occurs, the standard dichromate titration using 0.1 N potassium dichromate with barium diphenyl sulfonate as indicator is sufficiently sensitive to determine iron content.

After the sample is dissolved in concentrated hydrochloric acid and diluted with distilled water, stannous chloride is used to reduce ferric iron to the Fe(II) state. Excess stannous chloride is precipitated by the addition of mercuric chloride.

Equally sensitive, titration with 0.01 N ceric sulfate is performed with endpoint detection using 0.025 N 1,10-phenanthrolene-ferrous complex.

Application of this determination can be made to airborne soil from dustfall as well as to thermal or electrostatic precipitator samples.

The lower detection limit allowable by this approach is 100 µg of iron. Possible interferences include fluoride and phosphates, which may complex the ceric ion.

REFERENCES

1. J. S. Amenta, "Spectrophotometric Determination of Carbon Monoxide in Blood," in *Standard Methods of Clinical Chemistry*, D. Seligson, ed., Vol. 4, Academic Press, New York, 1963, pp. 31–38.
2. L. Vignoli et al., "Observations on the Use of Reactive Tubes of the Draeger-type for the Determination of Blood Carbon Monoxide," *Arch. Mal. Prof. (Paris)*, **21**, 201–204 (April–May 1960).
3. R. F. Coburn, "Carbon Monoxide in Blood: Analytical Method and Sources of Error," *J. Appl. Physiol.*, **19**, 510–515 (1964).
4. K. Porter and D. H. Volman, "Flame Ionization Detection of CO for Gas Chromatic Analysis," *Anal. Chem.*, **24**, 748 (1962).
5. J. D. Hackney et al., "Rebreathing Estimate of Carbon Monoxide Hemoglobin," *Arch. Environ. Health*, **5**, (4), 300–307 (1962).
6. J. E. Peterson and R. O. Stewart, "Absorption and Elimination of Carbon Monoxide by Inactive Young Men," report prepared by the Coordinating Research Council Inc., Contract CRC-APRAC No. CAPM-3-68, 1969.
7. D. R. Bates and A. E. Witherspoon., "The Photochemistry of Some Minor Constituents of the Earth's Atmosphere (CO_2, CO, CH_4, NO_2)," *Monthly Notices Roy. Astron. Soc. (London)*, **112**, 101–124 (1952).
8. A. P. Altshuller and J. J. Bufalini, "Photochemical Aspects of Air Pollution: A Review," *Photochem. Photobiol.*, **4**, (2), 97–146 (1965).
9. *Methods of Air Sampling and Analysis*, A. C. Stern, Chairman, Intersociety Committee, Methods of Air Sampling and Analysis, American Public Health Association Publication, Washington, D.C., 1972, pp. 239–241.
10. S. S. Wilks, "Carbon Monoxide in Green Plants," *Science*, **129**, 964–966 (1959).
11. E. G. Barham, "Sephonophores and the Deep Scattering Layer," *Science*, **140**, 826–828 (1963).
12. V. Middelton et al., "Carbon Monoxide Accumulation in Closed Circle Anesthesis Systems," *Anesthesiology*, **26**, 715–719 (1965).
13. J. M. Colucci, C. R. Begeman, and J. H. Neff, "Carbon Monoxide Concentrations in Detroit, New York and Los Angeles Air," *Environ. Sci. Tech.*, **3**, 41–47 (1969).
14. R. S. Brief, A. R. Jones, and J. D. Yoder, "Lead Carbon Monoxide and Traffic; A Correlation Study," *J. Air Poll. Control Assoc.*, **10**, 304–388 (1960).
15. J. L. Locke and L. Herzberg, "The Absorption Due to Carbon Monoxide in the Infrared Solar Spectrum," *Can. J. Phys.*, **31**, 504–516 (1953).

16. *Air Quality Criteria for Carbon Monoxide,* U.S. Department of HEW, PHS, National Air Pollution Control Admin. Publ. AP-62, Sect. 4, p. 1.
17. P. R. Ehrlich and A. H. Ehrlich, *Population Resources Environment,* W. H. Freeman & Co., New York, 1970, p. 119.
18. A. J. Haagen-Smit and L. G. Wayne, "Atmospheric Reactions and Scavenging Processes," in *Air Pollution,* A. C. Stern, ed., Vol. 1, 2nd ed., Academic Press, New York, 1968, p. 181.
19. M. Stephenson, *Bacterial Metabolism,* 3rd ed., Longmans, Green and Co., New York, 1949, p. 398.
20. S. A. Waksman, *Principles of Soil Microbiology,* 2nd ed., Williams & Wilkins Co., Baltimore, January 1932, p. 894.
21. M. J. Shepherd, "Modifications of Apparatus for Volumetric Gas Analysis," *J. Res. Natl. Bur. Stand.,* **21,** 351 (1941).
22. E. G. Adams and N. T. Simmons, "The Determination of Carbon Monoxide by Means of Iodine Pentoxide," *J. Appl. Chem. (London),* **1,** S20 (1941).
23. P. Hersch and C. J. Sambucetti, Pittsburgh Conference Analytical Chemistry and Applied Spectrometry, 1963, as reported by M. Katz, *Air Pollution,* Vol. 2, A. C. Stern, ed., Academic Press, New York, 1968, p. 106.
24. G. Ciuhandu, "New Colorimetric Method for Determination of Carbon Monoxide in Air," *Acad. Rep. Populare Romine Studdi Cercetari Chim;* French Summary, *C.A.,* **51,** 4877 (1957), as reported in *Chemical Detection of Gaseous Pollutants,* W. E. Ruch, ed., Ann Arbor Science Publ., Ann Arbor, 1968, p. 68.
25. D. A. Levaggi and M. Feldstein, "The Colorimetric Determination of Low Concentrations of Carbon Monoxide," *Am. Ind. Hyg. Assoc. J.,* **25,** 64–66 (1964).
26. R. G. Smith et al., "Tentative Method of Analysis for Carbon Monoxide Content of the Atmosphere," *Health Lab. Sci.,* **7,** 75–77 (1970).
27. A. O. Beckman, J. D. McCullough, and R. A. Crane, "Microdetermination of Carbon Monoxide in Air, A Portable Instrument," *Anal. Chem.,* **20,** 674 (1948).
28. J. M. Salsburg, J. W. Cole, and J. H. Yoe, "Determination of Carbon Monoxide—A Microgravimetric Method," *Anal. Chem.,* **19,** 66–68 (1947).
29. I. Lysijj, J. E. Zarembo, and A. Hanley, "Rapid Method for Determination of Small Amounts of Carbon Monoxide in Gas Mixtures," *Anal. Chem.,* **31,** 902–904 (1959).
30. R. G. Smith et al., "Tentative Method of Analysis for Carbon Monoxide Content of the Atmosphere (Infrared Absorption Method)," *Health Lab. Sci.,* **7** (1), 78 (1970).
31. F. H. Goldman and A. D. Brandt, "A Comparison of the Methods for Carbon Monoxide," *Am. J. Publ. Health,* **32,** 475–480 (1942).
32. J. M. Salsbury, J. W. Cole, and J. H. Yoe, "Determination of Carbon Monoxide," *Anal. Chem.,* **19,** 66–68 (1947).
33. J. E. Johnson, J. G. Christian, and H. W. Carhart, "Hopcalite Catalyzed Combustion of Hydrocarbon Vapors at Low Concentrations," *Ind. Engr. Chem.,* **53,** 900–902 (1961).

34. R. G. Smith et al., "Tentative Method of Analysis for Carbon Monoxide Content of the Atmosphere (Hopcalite Method)," proposed May, 1974, Intersoc. Committee for Methods of Air Sampling and Analysis.
35. A. B. Lamb, W. C. Gray, and J. C. W. Frazier, "The Removal of Carbon Monoxide from Air," *Ind. Eng. Chem.*, **19**, 213–217 (1920).
36. M. Feldstein, *Progress in Chemical Toxicology*, A. Stoleman, ed., Vol. 3, Academic Press, New York, 1967, pp. 105–106.
37. R. G. Smith et al., "Tentative Method for Preparation of Carbon Monoxide Standard Mixtures," *Health Lab. Sci.*, **7**, 72–74 (1970).
38. M. Katz, "Inorganic Gaseous Pollutants," in *Air Pollution*, A. Stern, ed., Academic Press, New York, 1968, p. 108.
39. L. Silverman and G. R. Gardner, "Potassium Pallado Sulfite Method for Carbon Monoxide Detection," *Am. Ind. Hyg. Assoc. J.*, **26**, 97–105 (1965).
40. R. Leers, "Reactive Material for Carbon Monoxide Detecting Tubes," *C.A.*, **54**, 17991 (1960).
41. J. W. Cole, J. M. Salsbury, and J. H. Yoe, "Photoelectric Detection and Estimation of Carbon Monoxide with Granules Coated with Palladous Silicomolybdate," *Anal. Chem. Acta*, **2**, 115–126 (1948), as reported in *C.A.*, **42**, 8719 (1948).
42. F. A. Patty, ed. *Industrial Hygiene and Toxicology*, Vol. 2: *Toxicology*, D. W. Fassett and D. D. Irish, eds., Interscience, New York, 1963, pp. 893–894.
43. C. E. June and T. Ryan, "The Study of SO_2 Oxidation in Solution and Its Role in Atmospheric Chemistry," *Quart. J. Roy. Meteorol. Soc.*, **84**, 46–55 (1958).
44. E. R. Gerhard and G. F. Johnstone, "Photochemical Oxidation of Sulfur Dioxide in Air," *Ind. Eng. Chem.*, **47**, 972–976 (1955).
45. M. B. Jacobs, "Methods for the Differentiation of Sulfur-Bearing Components of Air Contaminants," in *Air Pollution*, L. C. McCabe, ed. New York, McGraw-Hill, 1952, pp. 201–209.
46. Air Quality Criteria for Sulfur Oxides, U.S. Dept. of HEW, PHS, NAPCO, Washington, D.C. Chapter 2, p. 20, 1969.
47. M. Katz, "Photoelectric Determination of Atmospheric Sulfur Dioxide, Employing Dilute Starch Iodine Solutions," *Anal. Chem.*, **22**, 1040–1047 (1950).
48. W. E. Middleton, *Vision Through the Atmosphere*, University of Toronto Press, Toronto, 1952, p. 250.
49. J. F. Roesler, H. S. R. Stevenson, and J. S. Nader, "Size Distribution of Sulfate Aerosols in the Ambient Air," *J. Air Poll. Control Assoc.*, **15**, 576–579 (1965).
50. J. Wagman, R. E. Lee, and C. J. Axt, "Influence of Some Atmospheric Variables or Concentration and Particle Size Distribution of Sulfate in Urban Air," *Atmos. Environ.*, **1**, 479–498 (1967).
51. P. W. West and G. C. Gaeke, "Fixation of Sulfur Dioxide as Sulfitomercurate III and Subsequent Colorimetric Determination," *Anal. Chem.*, **28**, 1816 (1956).
52. D. F. Adams et al., "Tentative Method for Sulfur Dioxide Content of the Atmosphere (Colorimetric)," *Health Lab. Sci.*, **7**, 5–12 (1970).

91. J. M. Heuss, G. J. Nebel, and J. M. Collucci, "National Air Quality Standards for Automotive Pollutants," *J. Air Poll. Control Assoc.,* **21,** 537 (1971).
92. G. B. Morgan, C. Golden, and E. C. Tabor, "New and Improved Procedures for Gas Sampling and Analysis," in the National Air Sampling Network.
93. C. M. Shy et al., "The Chattanooga School Children Study: Effects of Community Exposure to Nitrogen Dioxide," *J. Air Poll. Control Assoc.,* **20,** 539–548 (1970).
94. A. A. Christic, R. G. Lidzey, and D. W. F. Rudford, "Field Methods for the Determination of Nitrogen Dioxide in Air," *Analyst,* **95,** 519–524 (1970).
95. *Federal Register,* **38,** 15175 (1973).
96. *Federal Register,* **36,** 8201 (8.1.2) (1971).
97. D. L. Ripley, J. M. Clingenpeel, and R. W. Hurn, "Continuous Determinations of Nitrogen Oxides in Air and Exhaust Gases," *Int. J. Water Poll.,* **8,** 455–459 (1964).
98. *Federal Register,* **36,** 8300 (1971).
99. P. A. Leighton, *Photochemistry of Air Pollution,* Academic Press, New York, 1961, p. 300.
100. R. J. Cvetanovic, "Addition of Atoms to Olefins in the Gas Phase," in *Advances in Photochemistry,* W. A. Noyes et al., eds., Vol. 1, Interscience, New York, 1963, pp. 115–182.
101. A. P. Altshuller, "Photochemical Reactivities of *n*-Butane and Other Paraffines Hydrocarbons," *J. Air Poll. Control Assoc.,* **19,** 707–790 (1969).
102. E. R. Stephens, F. R. Burleson, and E. A. Cardiff, "The Production of Pure Peroxyacyl Nitrates," *J. Air Poll. Control Assoc.,* **87,** 87–89 (1965).
103. Air Quality Criteria for Photochemical Oxidants, U.S. Dept. of HEW, PHS, EPA, Publ. No. AP-63, 1970, Sect. 2, p. 10.
104. D. H. Byers and B. E. Saltzman, "Determination for Ozone in Air by Neutral and Alkaline Iodide Procedures," *J. Am. Ind. Hyg. Assoc.,* **19,** 251–257 (1958).
105. *Federal Register,* **36,** 1510 (1971).
106. *Federal Register,* **36,** 1512 (1971).
107. D. F. Adams et al., "Tentative Method for Manual Analysis of Oxidizing Substances in the Atmosphere," No. 44101-01-70T, *Health Lab. Sci.,* **7,** 152 (1970).
108. D. H. Byers and B. E. Saltzman, "Determination of Ozone in Air by Neutral and Alkaline Iodide Procedures," *J. Am. Ind. Hyg. Assoc.,* **19,** 251–257 (1958).
109. A. J. Haagen-Smit and M. F. Brunelle, "The Application of Phenolphthalin Reagent for Atmospheric Oxidant," *Int. J. Air Poll.,* **1,** 51 (1958).
110. J. J. Bufalini, "Gas Phase Titration of Atmospheric Ozone," *Environ. Sci. Tech.,* **2,** 703–704 (1968).
111. T. Vega and C. Seymore, "A Simplified Method for Determining Ozone Levels in Community Air Pollution Surveys," *J. Air Poll. Control Assoc.,* **11,** 28–33 (1961).

112. E. R. Stephens, F. R. Burleson, and E. A. Cardiff, "The Production of Pure Peroxyacyl Nitrates," *J. Air Poll. Control Assoc.,* **87,** 87–89 (1965).
113. A. P. Altshuller, "Reactivity of Organic Substances in Atmospheric Photo-oxidation Reactions," *Int. J. Air Water Poll.,* **10,** 713–733 (1966).
114. A. P. Altshuller et al., "Photochemical Reactivities of n-Butane and Other Paraffines Hydrocarbons," *J. Air Poll. Control Assoc.,* **19,** 787–790 (1969).
115. E. F. Darley, K. A. Kettner, and E. R. Stephens, "Analysis of Peroxyacyl Nitrates by Gas Chromatography with Electron Capture Detection," *Anal. Chem.,* **35,** 589–591 (1963).
116. R. G. Smith et al., "Tentative Method of Analysis for PAN Content of the Atmosphere," proposed Dec. 30, 1968; revised Aug. 1, 1972, Intersoc. Committee for Methods of Air Sampling and Analysis.
117. E. R. Stephens, F. R. Burleson, and E. A. Cardiff, "The Production of Pure Peroxyacyl Nitrates," *J. Air Poll. Control Assoc.,* **15,** 87–89 (1965).
118. E. R. Stephens, "Absorptivities for Infrared Determination of Peroxyacyl Nitrates," *Anal. Chem.,* **36,** 928–929 (1964).
119. L. V. Cralley et al., "Tentative Method of Analysis for Chloride Content of the Atmosphere," No. 22203-01-68T, *Health Lab. Sci.,* **6,** 61–63 (1969).
120. M. J. Price and O. J. Walker, "Determination of Fluoride in Water by the Aluminum Hemotoxylin Method," *Anal. Chem.,* **24,** 1593–1595 (1952).
121. L. V. Cralley et al., "Tentative Methods of Analysis for Fluoride Content of the Atmosphere and Plant Tissues (Manual Methods)," *Health Lab. Sci.,* **6,** 64–83 (1969).
122. J. P. Nielsen and A. D. Dangerfield, "Use of Ion Exchange Resins for Determination of Atmospheric Fluorides," *AMA, Arch. Ind. Health,* **11,** 61 (1955).
123. R. J. Rowley and G. H. Churchill, "Titration of Fluorene in Aqueous Solutions," *Ind. Eng. Chem., Anal. Ed.,* **9,** 551–554 (1937).
124. *Federal Register,* **36,** 23239–23256 (1971).
125. K. Magos, "Mercury Blood Interaction and Mercury Uptake by the Brain After Vapor Exposure," *Environ. Res.,* **1,** 323–337 (1967).
126. *Federal Register,* **36,** Part 61, 23239 (1971).
127. G. W. Wright, "Asbestos and Health in 1969," *Am. Rev. Resp. Dis.,* **100,** 467–479 (1969).
128. State of Illinois Pollution Control Board Newsletter, No. 40, Part V, Sect. 501, p. 6 (January 10, 1972).
129. M. Eisenbud et al., "Non-Occupational Berylliosis," *J. Ind. Hyg. Toxicol.,* **31,** 282–294 (1949).
130. H. E. Stokinger, "Recommended Hygienic Limits of Exposure to Beryllium," in *Beryllium: Its Industrial Hygiene Aspects,* H. E. Stokinger, ed., Academic Press, New York, 1966, p. 236.
131. S. D. Shearer, G. G. Akland, D. H. Fair, T. B. McMullen, and E. C. Tabor, "Concentrations of Particulate Lead in the Ambient Air of the United States,"

statement presented at Public Hearing of Gasoline Additive Regulations, Los Angeles, Calif., May 2-4, 1972.
132. S. Hernberg, J. Nikkanen, G. Mellin, and H. Lilius, "S-Aminolevulinic Acid Dehydrase as a Measure of Lead Exposure," *Arch. Environ. Health,* **21,** 140-145 (1970).
133. *Federal Register,* **36,** 2550 (1971).
134. G. Ter Haar and R. Aronow, "New Information on Lead in Dirt and Dust as Related to the Childhood Lead Problem," *Environ. Health Perspec.,* Exp. Issue No. 7, May 1974.
135. H. Weisz, "Explanation of Separations in a Drop," *Microchim. Acta,* 140-144 (1954).
136. P. W. West et al., "Transfer Concentration and Analysis of Collected Air-borne Particulates Based on Ring Oven Techniques," *Anal. Chem.,* **32,** 943-948 (1960).
137. P. W. West, "Chemical Analysis of Inorganic Pollutants," in Air Pollution, Vol. 2, A. C. Stern, ed., Academic Press, New York, 1968, pp. 160-164.
138. P. W. West and F. E. Ordovesa, "The Elimination of Nitrogen Dioxide Interference in the Determination of Sulfur Dioxide," *Anal. Chim. Acta,* **30,** 227-231 (1964).
139. G. D. Carlson and W. E. Black, "Determination of Trace Quantities of Metals from Filtered Air Samples by Atomic Absorption Spectrophotometry," presented at the 63rd Annual Meeting of the Air Pollution Control Assoc., St. Louis, Mo., June 14-18, 1970.
140. "Annual Report, 1970-71," Wayne County Air Pollution Control Agency, Detroit, Mi.
141. *Federal Register,* **36,** 23251 (1971).
142. E. C. Tabor et al., "Tentative Method of Analysis for Selenium Content of Atmospheric Particulate Matter," *Health Lab. Sci.,* **7** (1), 96-101 (1970).
143. E. C. Tabor et al., "Tentative Method of Analysis for Antimony Content of the Atmosphere," *Health Lab. Sci.,* **7** (1), 92-95 (1970).
144. E. C. Tabor et al., "Tentative Method of Analysis for Manganese Content of Atmospheric Particulate Matter," *Health Lab. Sci.,* **7** (3), 146-148 (1970).
145. E. C. Tabor et al., "Tentative Method of Analysis for Molybdenum Content of Atmospheric Particulate Matter," *Health Lab. Sci.,* **7** (3), 149-151 (1970).
146. E. C. Tabor et al., "Tentative Method of Analysis for Arsenic Content of Atmospheric Particulate Matter," *Health Lab. Sci.,* **6** (2), 57-60 (1969).
147. W. C. Fredrick, "Tentative Method of Analysis for Iron Content of Atmospheric Particulate Matter (Volumetric Method)," *Health Lab. Sci.,* **8** (1), 56-58 (1971).

4

CONTINUOUS, AUTOMATED METHODS OF AIR ANALYSIS

INTRODUCTION

In the urban environment, pollutant levels measured by collecting air for a few hours or for 1 hr periods scattered randomly over several days may not always present a true picture of the actual ambient air quality.

Since emissions of contaminants from urban industrial sources seldom occur at a constant rate, they rarely produce a uniform average air concentration. The air analyst who is equipped only with filters and test tube bubblers often must be satisfied with data that are based on average concentrations obtained by random and sometimes quite infrequent sampling of complex industrial sources.

In response to the need for a true picture of ambient air quality in recent years, progress in automating air sampling methods of analysis has led to the development of systems that continuously mix reagents, photometrically irradiate sample air streams, or weigh particulate. Many of these sampling units have proved to be only complicated extensions of the single-batch sampling methods in which manual zeroing, addition of reagent, or adjustment of sample flow rate have all too frequently erased the advantages of the so-called continuous monitor over the manually collected and analyzed air sample. In other instances, for example, automated determination of fluorescent or thermally reactive compounds whose analytical chemistry has already required rather sophisticated electronic hardware, application of continuous end-point detection has actually simplified the otherwise complicated field sampling approach by eliminating the need for sample preparation and separation steps. With the application of continuous flame photometry, hydrocarbons no longer need be extracted from glass-fiber filters for individual weighing, but may be introduced directly into a hydrogen–air flame whose change in temperature is rather simply chart recorded.

As an example of a recently developed method, single sample analysis for fluoride formerly required several separation steps to free the fluoride from such interferences as phosphorus and sulfur. However, a number of automated methods have been developed for continuous measurement of volatile fluorides by taking advantage of the fluoride ion as an agent in fluorescent quenching. Instrumentation here involves simply the use of a paper tape impregnated with alcoholic magnesium oxinate, whose fluorescence at 3650 Å is reduced in proportion to the concentration of volatile fluoride in the sample air that is continuously passed through the tape. A photocell detector is used to monitor levels of ambient fluoride to as low as a few parts per billion.[1]

Similarly, the application of chemiluminescence to continuous monitoring methodology has resulted in the development of instruments selective to nitrogen dioxide and to ozone. In the case of ozone monitoring, this approach has proved much simpler in unit operation than the continuous coulometric method employed by Regener,[2] which is based on liberation of iodine in the ozone–iodide reaction. Thus far, with only limited application in the field, the chemiluminescence approach has proven a reliable substitute[3] for the reference bubbler method,[4] which like the Regener method relies on the iodide–oxone reaction (in neutral buffered solution) to produce iodine whose absorbance at 352 mμ yields ozone concentration through the use of standard curves developed from iodine equivalents of O_3. Chemiluminescent determination of NO_2 is accomplished by the Nederbracht reaction in which NO_2 is first reduced to NO by passing through a heated stainless steel tube. The NO is then reacted with excess ozone to yield NO_2^* (430, 470 mμ).[5] This single measurement yields total nitrogen oxides (NO_x), where measurement of NO_2 arises from the difference between NO_x and NO ($NO_x^- = NO_2$).

Currently, continuous analyzers are available for the measurement of SO_2, H_2S, total sulfur compounds, CO_2, and even suspended particulate. Among the analytical principles employed are electrical conductivity, potentiometry, colorimetry, infrared and uv absorption, and the piezoelectric effect. Such instruments are frequently programmed with automatic zero, span (100%), and calibration cycles.

In general, continuous instrumentation is composed of separate functional units along the lines of the diagram in Figure 1.

GENERAL PERFORMANCE CHARACTERISTICS OF CONTINUOUS SAMPLING INSTRUMENTS

Using the diagram in Figure 1 as a guide, the portion marked absorber column indicates a chamber or scrubber containing the reagent solution (e.g., acid, base, or complex ion) required to effectively contact and retain the desired air contaminant in solution.

The partition equation, which represents contact time, is

$$W_n = w\left(\frac{KV_1}{KV_1 + V_2}\right)^n$$

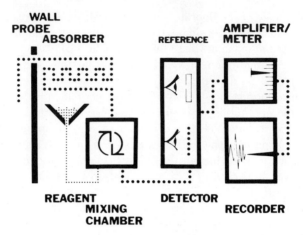

Figure 1. Schematic representation of a generalized autoanalyzer. (Published in the 1973 Institute of Environmental Sciences Proceedings, pages 239–245)

This equation uses the factor n to indicate the number of "contact times" needed to remove a given weight of material $(W - W_n)$. Such a factor becomes a function of the total retention time in the absorber. Although a long retention time may improve the overall absorbing effectiveness, such effectiveness is accomplished at the expense of response time, which is the difference between the time t_0 when a given parcel of air enters the sampling probe and the time t_r when the instrument recorder first begins to respond to the level of contaminant.

In the computing of average concentrations of pollutant, the actual response time makes very little difference for continuous samplers. It is only in cases of air pollution episodes or alert situations that an instrument operator must remember that the readout he observes may well be the product of a parcel of contaminated air that entered the instrument as much as 20 to 40 min prior to that time. In normal urban air monitoring with fluctuating air quality, a response time of 30 sec to 15 min is generally considered acceptable.

Sensitivity

Sensitivity is defined as the smallest concentration of single contaminant producing an instrument recorder reading that can be distinguished from the noise level of the instrument.

Instrument manufacturers often report the sensitivity at 50% of scale reading. This serves to further distinguish sensitivity from the "detection limit," which ordinarily is the sensitivity at the lowest concentration of pollutant distinguishable above zero scale reading.

Drift

Zero drift represents the change in instrument reading with time when clean air is passed through the sample portion of the instrument. Zero drift is usually expressed in percent of scale (e.g., 1, 2%/hr). This parameter yields a check on the reproducibility of the electronic circuits. A calibration check using a known concentration of pollutant may also reveal a drift at a reading that falls between 55 and 75% of scale.

Reliability

Frequency of repairs, and thus the length of instrument "downtime," generally reflects the complexity of the continuous analyzer. Common causes of such breakdowns include the following:

1. Plumbing joints, which permit clogging and crystallization of reagent solution.
2. Filters, which require periodic replacement.
3. Sensitivity of electronics and absorber medium to temperature change.
4. Too rigid construction of the solution flow system, where normal field handling and vibration experienced in transport tend to loosen or crack flow-system parts.

CONTINUOUS GAS ANALYZERS

In general, these instruments are based on one of the following principles of operation: colorimetry, atomic or molecular absorption, chemiluminescence, conductivity, coulometry, or combustion.

In the past, colorimetric instruments have been used with varying degrees of success to monitor air by adapting classical color-forming reactions to such plumbing and electronics as were required to produce continuous recorded data. More recently, however, the realm of solid-state physics has produced gas-sensing equipment that respond to physical rather than chemical properties at even the lowest levels of gaseous air contaminants.

Therefore, emphasis is placed on the more recent physical instrumentation for the individual air contaminants. Future development in continuous air monitoring systems will probably be along the lines of physics rather than solution or "chemical" measurement.

CO

Automated continuous methods for CO include applications of gas chromatography, nondispersive infrared absorption, catalytic oxidation, and displacement of Hg from HgO to produce mercury vapor.

Nondispersive Infrared Analyzer

The most commonly used instruments are those for CO measurement using the principle of nondispersive infrared, employing either a long path (40 in.) or, more recently, a 10-cm path of infrared radiation.

Principle of Operation

These analyzers depend on the characteristic energy of absorption of the CO molecule at not only its absorption wavelength maximum of 4.6 μ but also at a number of equally specific lines ranging from 2 to 15 μ, which together differentiate CO from such interferences as CO_2, H_2O, SO_2, and NO_2.

The differentiation of CO from other interfering gases depends on the overall difference in energy absorbed as infrared radiation is passed through (1) a sample cell containing CO, and (2) a sealed reference cell containing the major interference gases, CO_2, and water vapor.

As shown in Figure 2, these instruments employ a heated filament as the

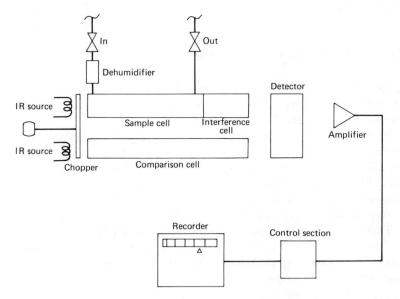

Figure 2. Diagram of nondispersive infrared analyzer. [Reprinted by permission from American Public Health Association, *Health Lab. Sci.,* **7** (1) 84 (1970)].

source of radiation, a chopper to alternate radiation between the sample and reference cells, a sample cell (usually copper or brass), a reference cell of the same material, and a detector.

Irradiation of the sample and reference cells (see Figure 1) produces heat in proportion to the number of CO molecules present to absorb heat energy. This causes pressure changes in the detector, which are sensed as electrical signals. A very low flow of dried sample air, approximately 150 ml/min, passes through the sample cell, while the reference cell maintains a fixed quantity of nitrogen, CO, and water vapor.

As infrared radiation is absorbed by the sample air in the sample cell, captive air in the detector unit contracts in response to a decrease in energy in proportion to the part per million concentration of energy-absorbing CO contained in the sample air. Although other gases (e.g., CO_2, N_2, H_2O, and O_2) absorb in the infrared as well, these gases are present in concentrations normally found in ambient air, as reference gas, in the reference cell.

These instruments operate at atmospheric pressure and are calibrated simply by passing a known part per million concentration of CO in nitrogen into the sample cell with about 1 to 5 min allowed for equilibration. The zero is set using pure nitrogen.

As one might expect, the longer the cell path, the greater the achievable sensitivity. This principle holds up to a point where extreme cell length (100 m) adversely affects the tuning of the electronic output of the detector, making an ultraextended cell length approach impractical. Ordinarily, instruments of approximately 1-m cell length are able to measure from 1 to 50 ppm. However, recently introduced instruments using cell paths as short as 10 cm have proven capable of measuring 1 to 25 ppm. Here, an advantage is gained in response time when a shorter cell path is used as long as flow rate remains constant. Depending on this cell length factor, the response time of most instruments varies from 1 to 5 min.

Interferences

Interferences, notably water vapor, have been observed to adversely affect the continuous determination of CO in spite of attempts to maintain conditions of high relative humidity in the reference cell as a compensation factor. Recently, optical filters have been placed ahead of the sample cell to minimize this interference by limiting the IR radiation window to a range in which neither CO_2 nor water vapor absorbs infrared energy. In this respect, it has been the experience of the author that, especially in summer months, the relative humidity maintained in the reference cell is very often too low to zero out the significant effect of water vapor in warmer ambient air.

Evaluation of data from the U.S. Weather Bureau has indicated that CO readings obtained in a tropical climate may be as much as 10 ppm higher than those obtained in dryer climates. Less interference would be expected in temperate zones, especially during the winter months when absolute humidity is lower.

While drying of the sample air using such desiccants as calcium sulfate has proven rather unsatisfactory,[6] drying by means of freezing coils has proven helpful in cases where ambient humidity contributed to the background output of even those NDIR analyzers equipped with optical filters.

Advantages of this system include its nonreliance on wet chemicals and ease of operation to the point where nontechnical personnel can serve as operators. These instruments are also rather insensitive to temperature change and are often equipped with automatic zeroing, spanning, and calibration systems.

Reagents Required

1. Span gas—A mixture of CO in nitrogen at a parts per million concentration approximately two-thirds of the full-scale range selected.

2. Zero gas—Pure nitrogen or a low-CO nitrogen containing less than 5 ppm CO of concentration known to the nearest $+1$ ppm. Zero gas may also be produced dynamically as the effluent from a catalytic oxidizer that converts CO to CO_2.

Procedure

Specific procedures are supplied in the manufacturer's operating manuals. However, the following general operating outline provides the reader with some degree of familiarization with general operative techniques.

1. Turn on power and allow analyzer to warm up for a period of 30 min to 3 hr.

2. Introduce zero gas and adjust to zero. Introduce span gas and adjust scale reading to the known part per million concentration of the span gas.

3. Adjust sample flow rate to specifications provided with instrument.

4. Introduce two to three known concentrations of CO in nitrogen, which may be obtained by diluting aliquots or the actual flow of the span gas with CO-free nitrogen. Prepare a calibration chart using the instrument response to these known CO concentrations.

5. Check for interferences by drawing a sample of air of known high (60 to 80%) relative humidity through the sample cell and observe any reading so obtained. If a known CO_2 atmosphere is available, an interference check should also be run to determine the extent of this interference. As a rough interference check, a bag of known volume containing room air can be saturated with water to produce a known humidity after 1 to 2 hr equilibration at room temperature. Here, the response of the instrument is assessed by calculating the volume part per million concentration of water necessary to produce a 1 ppm response as CO.

Calculations

When a linear calibration curve results in the range of 0 to 100 ppm and no significant interference appears associated with water vapor, readings can be made directly from chart paper. However, it is common to find a baseline or interference drift that appears in the zero mode of operation.

The NDIR method for CO is recognized as a tentative method (Method 42101-04-69T)[7] of the Intersociety Committee for Methods of Ambient Air Sampling and Analysis.

It would be worthwhile here to note that a single-sample IR method for CO is also recognized by the Intersociety Committee (Method 42101-03-69T); however, restriction of this method to an IR equipped with a 10-m path length limits the practical applicability of the analysis to a laboratory having this rather exclusive IR cellpath. In regard to the rather wide acceptance of NDIR method, collaborative testing by Southwest Research Institute has shown that no significant interference was observed because of the effect of water vapor and overall evaluation produced evidence of repeatability and reproducibility within a 95% confidence interval.[8]

Maintenance

Maintenance ordinarily consists of periodic cleaning of the sample and cells, recharging the reference cell with water saturated CO_2, replacing span gas cylinder, attending to any sample air drying filter or refrigerated coils, and normal repairing and cleaning of the pump.

Gas Chromatographic Determination of CO by Reduction of CO to Methane

Principle of Operation

An air sample is introduced into a pretreatment or stripper column, which serves to retain the heavy hydrocarbons while passing CO and methane quantitatively into a chromatograph and then into a catalytic reducing chamber. The methane first passes through the reducing chamber unchanged, while the CO is reduced to methane. Both "methane" peaks are sensed by the hydrogen flame ionization detector. The response proportional to the first peak in time is attributed to methane while the second peak's response is integrated and counted as CO. Aside from possible instrumental failure to remove nonmethane hydrocarbons, the method has no known interferences. Accuracy is set at $+2\%$ of the absolute CO value, with a drift rarely observed greater than 1% of scale for a given 24-hr period.

The apparatus (described later in the analysis of nonmethane hydrocarbons) for this method is supplied by several instrument manufacturers. As shown in Figure 7, it consists of automatic switching valves, a time sequence programmer, a stripper column, a chromatographic column, a flame ionization detection column, and a catalytic reactor for converting CO to CH_4.

The catalytic reactor generally consists of approximately 6 in. of ¼- in. OD stainless steel tubing, which is packed with a mixture of nickel nitrate on 42 to 60 mesh firebrick that has been dried at 100°C while purging with a stream of dry nitrogen.

Operating Procedure

This analyzer can be operated on continuous, batchwise, or intermittent modes of sample flow. Ordinarily, sample air flows through the sample loop at 100 ml/min. For "automatic batch analysis," once every 5 to 10 min a sample is injected into the analyzer from the sample loop.

Alternatively, sample air can be accumulated in an integrating reservoir, such as a 10 to 20 liter inert plastic bag, so that the sample introduced from this bag into the stripper column represents an average part per million concentration for the time period over which ambient sampling took place. This allows fewer injections into the chromatograph.

Samples may also be injected manually using 15-ml gas aliquots, which are passed directly into the sample loop. An electronic chart readout yields part per million concentration directly from the integral CH_4 peak area when a calibration curve is prepared from at least three known concentrations of CO in nitrogen or ultrapure air.

Calculations

In general, peak heights of CO and CH_4 are sufficient to provide quantitative part per million concentrations where the 0 to 1 ppm range yields a sensitivity of approximately 50 ppb (0.050 ppm). Further references to the applicability of this method can be found in the works of Stevens, O'Keefe, and Ortman.[9]

The Continuous Hopcalite Method

Principle of Operation

This approach depends on the passage of air containing CO over an oxidation catalyst, Hopcalite, with CO being converted to CO_2 in the presence of air oxygen. Subsequent monitoring of the temperature of the exit air stream affords a thermal measure proportional to the concentration of CO, whose heat of oxidation reaction serves to elevate the exit air stream temperature.

Use of this method for detection of high concentrations of CO dates back to 1920 with the work of Lamb et al.[10] Later investigations of reaction conditions sample treatment, and catalyst conditioning parameters[11] have produced a reliable method of analysis for continuous or batchwise determination of ambient CO.

Since this method is best employed to measure concentrations of CO in excess of 50 ppm to an accuracy of ±5 ppm, its application remains limited to moderate to high CO concentrations (although commercially available analyzers are normally graduated to a scale range of 0 to 500 ppm). Consequently, the NDIR or gas chromatographic methods are usually recommended for determination of CO in the 0 to 5 ppm range. Here, preference to the Hopcalite oxidation method would follow from the field method considerations such as ruggedness and relative ease of maintenance.

Continuous Measurement of CO by Displacement of Mercury Vapor

A very recent addition to the instrumentation available for CO is the application of a method that is based on the reduction of mercuric oxide by CO to produce mercury vapor, and CO_2 [the latter has been referred to (Chapter 3) as capable of measurement by the gravimetry of ascarite absorption]. In this approach, it is the mercury vapor that lends itself to trace measurement by UV absorption detection.

Principle of Operation

In principle, any mild reducing agent is capable of producing Hg vapor in this reaction. In theory, therefore, methane as well as hydrogen should be eliminated from the air sample for the proper Hg–CO stoichiometry to produce reliable results. Compensation is made for atmospheric hydrogen by zero adjustment. However, methane has not been found to interfere under specific reaction conditions of 210°C and dry sample air. Under these conditions, a contact time of only 1.5 msec is required.

One of the most recent instrument modifications employs a bed of silver oxide, which serves to aid zeroing out any H_2 interference effect. Here, Robbins et al. found Hg_2O to be insensitive to H_2 oxidation (with no interfering heat of reaction) at H_2 concentrations as high as 10 ppm.[12] A schematic of the apparatus is shown in Figure 3. The response time of the Hg vapor instrument is similar to that of the NDIR (i.e., in the range of 1 to 5 min). Its limitations lie chiefly in its electronic instability.[13] However, its wide range (0.05 to 10.0 ppm) and good sensitivity recommend its use in determining background levels of CO where the need for the real-time data is not a major factor in instrument selection.

Sulfur Dioxide (SO_2)

Among the earliest applications of continuous analyzers to ambient air monitoring were those involving measurement of SO_2.[14,15] Both continuous and intermittent (sequential) sampling methods have been employed. These often

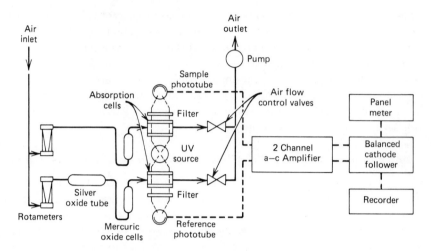

Figure 3. Schematic of Stanford Research Institute Carbon Monoxide Analyzer which employs Ag_2O and HgO oxidizing beds. (From reference 12.)

made use of the colorimetric method of West and Gaeke, as discussed in Chapter 3. It should be noted here that the West–Gaeke method was first adopted as the approved reference method by the National Air Pollution Control Association,[16] before being replaced by the EPA colorimetric method. Historically, if one excludes the various adaptations of continuous methods involving the ability of SO_2 to reduce starch–iodine solutions, the next most recent approaches to continuous monitoring make use of conductimetry in some form. Classically, the basis for these methods is the oxidation of SO_2 to sulfuric acid using aqueous dilute hydrogen peroxide, with subsequent measurement of the increase in electrical conductivity of the absorbing solution. Because of the less than 100% efficiency with which SO_2 is absorbed by water, the oxidation step is necessary to convert the gas to its soluble anionic form.

However, because the property of electrical conductivity is itself quite unselective, interferences in this method may well be expected from any strongly water-soluble gas (e.g., vapors of chlorine or hydrogen chloride) as well as from ionic particulate (e.g., salt spray), or even such weakly ionizing or less-soluble gases as H_2S or NO_2. Other interferences that are encountered less frequently include sulfuric acid mist, which produces a positive reading, and ammonia or amines. The latter can result in a negative interference by abstracting conductive protons from the system to form ammonium ion, thereby essentially neutralizing the conductive effect of absorbed SO_2. Such particulate interferences can be excluded from reaction by placing a glass-fiber filter or filter plug, such as glass wool, ahead of the sample probe. However, such a practice may, after several months exposure, result in absorption of ambient hydrocarbons on the inside walls of the probe as a layer that will itself act to absorb SO_2.

In recent years, a method widely employed in Europe to measure SO_2 employs aqueous 0.03 N hydrogen peroxide at a pH of 5 to collect the gas by means of a bubbler apparatus, with subsequent automatic titration of the sulfuric acid formed using standard alkali. Needless to say, this method is quite adaptable to the conductimetric end point where both means of end-point detection are widely used outside of the Western Hemisphere.

Comparisons of the West–Gaeke method with various continuous peroxide and conductivity measurements for SO_2 have yielded no overall correlation,[17] although Booras and Zimmer noted that the conductimetric approach usually resulted in higher values.[18]

Coulometry

Principle of Operation

Since 1960, several commercial instruments have employed iodine coulometry in aqueous solution to monitor SO_2 through the ability of SO_2 to reduce I_2, which is maintained in aqueous buffered equilibrium with iodide (I^-) through the following ionic reaction.

$$I_2 \pm SO_2 + 2H_2O \rightarrow SO_4^{-2} + 2I^- + 4H^+$$

The coulometry of the system requires two platinum electrodes (as anode and cathode). These cathodes maintain a trace of I_2 in equilibrium with 0.0002 M potassium iodide, which contains 1.5 M potassium bromide to maintain conductance properties. A third reference electrode is used to sense a shift in the anode-cathode potential during absorption of SO_2 from sample air, which is bubbled through the solution above in a detector cell.

Maintenance of constant sample air flow is absolutely necessary, since the measured air concentration of SO_2 is a function only of the mass of I_2 reacted per unit time during any given interval of time. In addition, solution factors (e.g., accumulation of sulfate ions) together with buffer deterioration after prolonged operation contribute to erroneous readings, drift, and slow response time. Several commercial manufacturers produce this buffered absorbing reagent. However, a suitable substitute can be prepared by mixing:

0.332_g KI
178.0_g KBr
13.8_g Na H_2 PO_4 H_2O
14.2_g Na_2 HPO_4 (anhydrous)

and dissolving to form 1 liter of solution.

Detection and Sensitivity

In practice, the coulometric approach yields a detection limit of 0.01 ppm, although manufacturers generally specify 0.02 ppm minimum detection limit. The ultimate sensitivity of the method is 10% of the lowest instrument range or 0.01 ppm.

Since the coulometric electronic response is not instantaneous, about 4 min normally must be allowed to produce 90% of the ultimate signal for any measurable concentration of SO_2.

Maintenance

Maintenance of this equipment usually involves replacement of the reference electrode every 6 to 12 months. It is also necessary to return the absorbing cells to the laboratory every 3 months for cleaning of the platinum electrodes in aqua regia and replacement of the buffered potassium iodide absorbing solution.

Preventive maintenance includes twice weekly flushing of the capillary tube that conveys sample air to the absorbing cell to dissolve small crystals that tend to block the entrance to the absorbing cell.

Zeroing of the instrument is achieved by setting the three-way intake valve to pass ambient air through a granular carbon scrubber, which removes all SO_2 and interferences. The resulting air is the referenced as "zero." Sample air is produced by switching the valve selector to the "sample" mode in which ambient air is passed through a selective scrubber, which removes such interferences as H_2S, halogen gases, and ammonia while passing SO_2.

Usually, a 15 to 30 min period of "zero" once per day is sufficient to allow for variation in response time to permit the setting of a true zero. Preventive maintenance also includes replacement or thorough cleaning of the sample probe (usually glass or Teflon) at least once every 6 months to prevent a buildup of hydrocarbons, which absorb ambient SO_2. Lack of such maintenance may result in as much as a 30 to 50% positive error after 9 to 12 months of continuous operation.

A number of instruments marketed only recently are equipped with a device for automatic water addition. This has eliminated the necessity to restore evaporated moisture by daily water addition, an advantage that has added measurably to the ease of operation of such continuous air monitors.

Flame Photometric Continuous Analyzer

For the past several years, the monitoring of sources such as kraft paper mills and oil refineries, whose emission requires a continuous total sulfur analyzer, has been accomplished by means of a total combined-sulfur flame photometer.

Principle of Operation

In this analyzer, sample air is admitted into a hydrogen-rich air flame. Specificity to sulfur arises from the use of a narrowband interference filter that shields the photomultiplier tube detector from all but the 394 mμ emission energy of flame-excited sulfur atoms.

Absence of any reaction time lag, as is the case with the coulometric SO_2 analyzer, results in an essentially real time output of data, where actual measured response time is 2 to 3 sec.

Here, as with other flame photometers, baseline drift remains a problem. If available, a self-zeroing recorder can automatically compensate for drift. During these zeroing periods, SO_2-free air may be obtained as either compressed air already on hand for operation of the detector flame or as laboratory air purified using a packed-carbon zero gas filter.

Research has been undertaken to increase the specificity of the instrument to actual SO_2 rather than total sulfur monitoring. Suitable filter systems, such as silver compounds, remove such interferences as hydrogen sulfide, some organic sulfur, and mercaptans. It is safe to say that selective filtration will soon produce a flame photometer with adequate efficiency for monitoring of ambient SO_2.

Reliability

Recently, several field comparison studies of continuous SO_2 analyzers have been made.[19,20] In the findings of Stevens et al.,[20] good agreement was obtained between two analyzers equipped with flame photometric detectors (FPD). In one case, the FPD was equipped with a gas chromatograph to separate the influence of H_2S, from the total sulfur response. On the basis of this side-by-side differential evaluation, an estimated 90% of the total sulfur was judged to be SO_2.

Conclusions resulting from this study produced the following information:

1. Recorded levels produced from coulometric and conductimetric instruments were generally higher than simultaneous levels obtained using West–Gaeke or flame photometric methods.

2. Flame photometric, conductimetric, and coulometric analyzers generally required less maintenance than wet colorimetric analyzers.

3. Total sulfur in the area measured (Los Angeles) was found to be over 90% SO_2. Under these conditions, continuous flame photometry might be considered accurate to within 10% of the actual SO_2.

Sensitivity

Sensitivity to total sulfur is as low as 0.01 ppm, which compares favorably to that of the continuous coulometer.

Nitrogen Oxides

Nitrogen Dioxide (NO_2)

Colorimetric Method

Until quite recently, continuous analyzers for NO_2 have employed the Griess–Saltzman method (described in Chapter 3) of a similar modified colorimetric method.

Since NO_2 is somewhat difficult to absorb in aqueous solution, an efficient absorbing column must be employed to continuously extract NO_2 from sample air passing into the analyzer. A 15-in. column made from ¼-in. ID glass tubing wound into a helix has been found to approach 100% collection efficiency, although a 30-in. column is sometimes preferred to ensure reliability. In addition, as in the case in the manual Saltzman methed, approximately 15 min must be allowed between initial contact of the color-forming reagent with the ambient air stream and final reading of the absorbance of the colored-dye. This means that the flow of reagent and sample air, which meet at the entrance of the absorbing column, must be slow enough so that at least 15-min will have elapsed before the absorbing solution containing a parcel of air contacted at some time, t_0, is presented, color-formed, to the detector cell.

This time delay is built into most commercial instruments by means of extended plumbing and slow reagent flow ahead of the colorimeter, which usually employs a 20 to 40 mm bandpass filter to measure color absorbance in the red region at 550 mm.

Colored, although not necessarily exhausted, reagent is usually discarded continuously. By using a recorder equipped with a "floating" zero, the color of this absorbing solution can be "zeroed out" or read as "zero" enabling continuous reuse of reagent.

Sensitivity of 0.01 ppm is obtained when the instrument is properly calibrated either chemically with standardized sodium nitrite solution (Chapter 3) or dynamically with NO_2 permeation tube (Chapter 6).

Chemiluminescence Method

Recently, several continuous NO_2-measuring instruments operating on the principle of chemiluminescence have been marketed. Here, a photomultiplier detector is used to measure the luminescence produced in the gas phase reaction between ozone and NO. Self-absorption of luminescence is minimized by placing the detector window directly against the face of the narrow gas mixing chamber. A broad band of radiation between 0.6 and 3μ is produced in this completely dry reaction:

$$NO + O_3 \rightarrow NO_2^* \rightarrow NO + h\nu$$

from which energy is emitted when the electronically excited NO_2^* molecules revert to the ground state, NO_2.[21]

Obviously, this method directly measures NO rather than NO_2. It is mentioned here because it forms the basis for a reliable differential measurement of NO_2 through the use of a reducing medium such as stainless steel at 110°C, to convert NO_2 to NO. Subsequent reaction of NO, thus formed, with ozone produces chemiluminescence equivalent to NO_x, where $NO_2 = NO_x - NO$. The sensitivity of this method is reported as 0.01 ppm. To date, sufficient field experience has been obtained to indicate the overall reliability of the instrument over long periods of operation.

NO

The only available direct measurement of NO presently available as a practical field method is the chemiluminescence method. Alternatively, an oxidizing scrubber may be used to convert NO to NO_2, with colorimetric measurement of total nitrogen oxides (NO_x) employed to yield NO concentration differentially by $NO = NO_x - NO_2$.

Such scrubbers are usually prepared by soaking strips of filter paper in potassium dichromate or potassium permanganate solutions to form media that can be used as dry, loosely packed oxidizing filters.

Generally, these filters, although convenient to implant in an air flow line ahead of the absorber, exhibit oxidizing efficiencies and overall lifetime that are highly dependent on relative humidity.[22] Probably the best choice of oxidizer for the differential continuous colorimetric measurement of NO is a 1% aqueous potassium permanganate solution through which sample air is bubbled. Here, 10 ml of this solution are placed in a fritted bubbler, followed successively by an empty impinger that acts as a trap, and a bubbler containing Saltzman absorbing solution. Maintenance procedures require changing of the oxidizing solution about once per day and cleaning of the fitted glass bubbler to prevent buildup of manganese dioxide in the pores of the frit.

Total Nitrogen Oxides (NO_x)

Using any of the previously mentioned methods for NO_2, a scrubber as previously described, must be included to effectively oxidize all ambient NO and N_2O for measurement as the total NO_x.

Using the chemiluminescent measurement for NO, the air sample is passed through a heated stainless steel tube in order to reduce NO_2 to NO. Total NO_x then follows directly from the measurement of total NO.

Ozone (Oxidant)

Although ozone is the specific allotrope of oxygen that occurs on addition of energy to O_2 under specific reaction conditions, "oxidant" can be defined as any of the group of ozonelike compounds that will oxidize a reference material or solution.

Oxidant

The two classical continuous methods for oxidant analysis are based on:

1. The red color or iodine formed when atmospheric oxidant reacts with a solution of potassium iodide (colorimetric analysis).
2. The measurable amperage produced when iodine is released from the aqueous reaction of atmospheric oxidants with a potassium iodide solution, (coulometric analysis).

Since selectivity depends only on the oxidizing ability of the pollutant, any reducing agents (e.g., SO_2 or H_2S) present in the air would exhibit a negative interference on either of these reactions. Consequently, by definition, the mere passage of ambient air through the KI solution produces net detector response and "net oxidant" value. Therefore, greater selectivity to true ozone is achieved by interposing, ahead of the absorbing solutions, an oxidizing scrubber (described in Chapter 3) to convert SO_2 into nonreactive sulfate ions and nitrogen oxides into subtractable NO_x.

Continuous Colorimetric Analysis

Analysis by continuous colorimetry is most often accomplished by passing sample air at a known rate through a solution of 1, 10, or 20% neutral buffered potassium iodide. Oxidants in the sample air react to produce red "free iodine," which is sensed by passing the solution through a photometric detector cell. This cell measures the difference in absorbance at 352 nm between exposed and unexposed KI solution, and the corresponding voltage output, calibrated to known iodine or ozone standards, is transferred to a strip chart recorder.

This continuous method is extremely sensitive and is quite suitable for measurement of as low as 0.005 ppm oxidant.

Continuous Coulometric Analysis

The coulometric or Mast method is also widely used as a method for continuous air analysis.[23] Here, the external current flow produced in response to the reaction

$$2I^- + \text{oxidant} \rightarrow I_2 + \text{ions}$$

causes a change in the output of a microammeter, which has been calibrated using known concentrations of ozone.

In this electrometric procedure, a small initial potential of 0.025 to 0.3 V is applied between the anode and cathode of the detector cell to consume a small amount of iodide ion and form a small amount of free iodine. This establishes a polarization equilibrium, which may be upset by absorption of sample air containing oxidizing (or reducing) species. As sample air bubbles through the detector cell, oxidizing compounds (especially ozone) consume iodide ions, causing a measurable flow of external current to reestablish polarization.

Sensitivity of the method is less than 0.01 ppm, while the best working range lies between 0.01 and 0.20 ppm ozone.[24] This range lies at the upper end of the sensitivity necessary to measure oxidant according to requirements set forth by the Environmental Protection Agency by whose authority an Ambient Air Quality Standard of 0.08 ozone annual mean has been established. Interferences include NO_2, H_2S, and SO_2 and are roughly the same as those for colorimetric method.

Comparison of the Colorimetric and Coulometric Methods

In the past few years, widespread use of both coulometric and colorimetric instruments have led to comparison of their analytical accuracy and field equivalence.

The work of Potter and Duckworth[25] indicates that field monitoring of ozone in atmospheres containing less than 0.20 ppm with corrections for NO_2 interference produced scale readouts that were not significantly different from those expected if two of the same kind of instrument of either type were operated side by side.

Similarly, the work of Cherniack and Bryan,[26] which includes measurement at levels above 0.20 ppm, indicates a correlation coefficient of 0.87 for colorimetric/coulometric readings obtained from the two instruments, which were calibrated from the same ozone source.

Probably the greatest differences observed between methods were found to relate to KI concentration differences between colorimetric instruments. Here, the higher the percentage of KI (10 to 20%), the greater the NO_2 interference[26,27] and the greater the possibility for side reactions.

It should be noted that the continuous colorimetric method that has been found by a number of researchers, including the Wayne County Air Pollution control laboratory, to give the least interferences and greatest reliability is the 1% neutral buffered KI method when equipped with a chromic oxide scrubber to eliminate SO_2 interference. Correction for NO_2 interference is made by making side-by-side NO_x measurements using the continuous colorimetric (Saltzman) method. This approach is specified by the Environmental Protection Agency (EPA).[28]

Calibration

Dynamic calibration can be achieved by passing charcoal-scrubbed air at a known constant rate through a chamber containing a controlled-potential uv lamp. When precautions, such as electronic stabilization of line voltage to maintain constant filament energy are taken, the part per million concentration of the effluent ozone reference mixture becomes a function only of the rate of air flow through the uv lamp chamber.

Techniques such as manual colorimetry or coulometry are then used to standardize the "ozone generator" by passing the effluent air mixture through a potassium iodide bubbler. Once standardized by dynamic calibration, an ozone generator can be used as a secondary standard from which to calibrate various total oxidant methods, which will then read out parts per million oxidant "as ozone." A diagram for such a generator as produced commercially is shown in Figure 4.

Ozone

Gas Titration Method

The almost instantaneous reaction of *trans*-1-2-butene with ozone gives rise to a quite specific measure of ozone in the presence of other oxidants. This

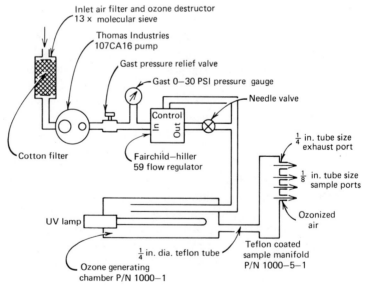

Figure 4. Airflow system for ozone generator. (Courtesy of McMillan Electronics, Corp., Austin, Texas.)

reactive butene also can be used to selectively gas-titrate ozone, while permitting the determination of the balance of total oxidants by standard continuous coulometric or colorimetric methods. This differential titration, described in Chapter 3 as a batch analysis method, can be transformed into an intermittent automatic method of air analysis by using electronic switching mechanisms. These mechanisims permit a measured volume of sample air to pass through the gas titration cell ahead of the colorimeter, while a second volume of sample air is introduced directly to the colorimeter. Here,

$$\text{ozone} = \text{total oxidant} - \text{ozonefree oxidant}$$

Chemiluminescence

The first chemiluminescent approach to a specific ozone determination probably was developed by Regener in 1960.[29] Regener found that, when air containing ozone contacts the surface of a plate prepared by absorbing rhodamine B on silica gel, a luminescence is produced from the chemical reaction. The intensity of the luminescence is proportional to the concentration of ozone present to concentrations as low as 0.001 ppm.

Regener's detector was found to be subject to a number of interferences, such as NO_2. It was soon followed by the Nederbracht detector, which employs the chemiluminescence of the ethylene reaction with ozone.[5]

A number of commercially available analyzers have now been marketed. It appears that the ozone–ethylene chemiluminescent reaction, having been adopted by the EPA as a standard method for ozone, will soon become the basis for the common continuous ozone field analyzer.

Since the chemiluminescent peaks produced are at approximately 430 and 470 mμ and since formaldehyde is spectrally active at 430 mμ, an aldehydic structure (—CHO), is suspected as being the product of this ozone ethylene reaction. Here, Pitts et al.[30] have suggested the *cis*-2-butene reaction:

It is quite possible that one or more such intermediates give rise to the observed chemiluminescence, and that any of a number of olefins could be employed as chemiluminescent initiators.

In practice, the continuous field instrument depends on the reaction that occurs when air and ethylene are drawn into a pyrex chamber and mixed directly in front of a pyrex window, behind which is mounted a photomultiplier tube. Any ozone present in the sample air reacts to produce a carbonyl intermediate plus a photon of chemiluminescent energy. This energy produces a signal from the photomultiplier tube in proportion to the total ozone present. Figure 5 shows schematically the air and ethylene reaction flow system.

In Figure 5, the entire photomultiplier tube assembly and electrometer system are contained in a housing that is cooled to $10° + 1°C$. This low temperature ensures insulation against change in ambient temperature, which would cause drift in the output of the photomultiplier tube.

Hydrocarbons

Commercial instruments that automatically measure hydrocarbons fall into two main categories:

1. The total hydrocarbon continuous monitor.
2. The semicontinuous nonmethane hydrocarbon monitor.

Briefly, automatic monitoring of hydrocarbon levels depends on the fact that most organic compounds easily pyrolyze when introduced into an air–hydrogen

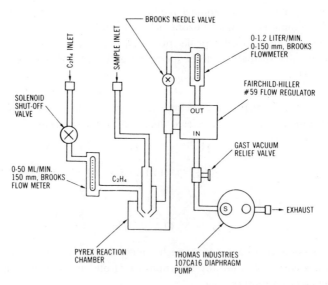

Figure 5. Diagram of Air-Ethylene System for Continuous Chemiluminescent Ozone Meter. (Courtesy of McMillan Electronics, Corp., Austin, Texas.)

flame. This pyrolysis produces ions that are collected either by the metal of the flame jet itself (charged negative) or by a cylindrical collecting grid (positively charged) that surrounds the flame. The sensitivity to organic materials varies slightly depending on the number and kind of ions. As a general rule, however, detector response is in proportion to the number of carbon atoms in the chain of the organic molecule. Thus propane (three carbon atoms) gives roughly three times the intensity of response as does methane, and so on.

This "nonselectivity" is both an advantage and a disadvantage, depending on the information expected from the air analysis. Nonselectivity toward kind of hydrocarbon, but selectivity in the sense that other compounds do not cause response, provides this continuous instrument with the capability of measuring the whole general class of organic compounds without concern for interference. When the instrument response is calibrated using methane, the continuous strip chart readout is then a record of the real-time variation in ambient hydrocarbons as though they were 100% methane.

Total Hydrocarbon Analyzer

Principle of Operation

In the case of most commercial instruments, the sampling system contains a sensitive flowmeter, which is designed to continuously measure the volume of ambient air sample that is mixed with hydrogen in the flame ionization detector. An electrometer coupled with a voltage reading recorder detects and charts the increase in ion intensity resulting from flame ionization of any organic compounds in the air sample.

The recording range of these analyzers varies from approximately 2 to 3000 ppm hydrocarbons as methane. Ordinarily, the 0 to 20 ppm range is most useful for atmospheric sampling, with an implied sensitivity of 1% full scale.

Interferences

Operationally, few interferences have been noted. These include compounds in which carbon is bonded directly to:

1. Oxygen (carbonyls).
2. Nitrogen (cyanides).
3. Halogens (halides).

These compounds produce a slight decrease in overall detector response. Here, of course, no detector response is observed in the case of CO, CO_2, or water vapor, in which no C—H bonds are found.

Calibration

Calibration of the instrument is accomplished by introducing zero gas (water-pumped compressed air containing less than 2 ppm hydrocarbon as methane) at the flow rate specified by the manufacturer and using the zero control to set the zero line or a desired scale value. (Here, a 10% offset of the "zero" up scale allows for possible negative scale drift. In this case, however, the span setting must also be offset by 10% up scale.)

It is often desirable to standardize the span gas against a known standard prepared from reagent hydrocarbons using a gas chromatograph and replicate samples to ensure a valid standard.

A schematic flow diagram of a commercially available continuous hydrocarbon analyzer is shown in Figure 6.

Some suggestions on the use of this type of continuous analyzer include the use of a suitable thermostat to ensure a constant temperature of instrument operation. Ordinarily, a range of $70 \pm 5°F$ is sufficient to stabilize the sample air–fuel flow system, which is quite sensitive to error from temperature expansion.

The Semicontinuous Nonmethane Hydrocarbon Analyzer

As mentioned earlier, the Federal ambient air quality standard of 0.24 ppm (6:00 to 9:00 A.M.) average for nonmethane hydrocarbons necessitates the selective measurement of this class of compounds in preference to total hydrocarbons, especially when elevated levels of ozone are either known or suspected (see Ozone, Chapter 3).

Principle of Operation

This analysis is accomplished by a differential measurement using the following procedure. First, small measured volumes of air are delivered intermittently (4 to 12 times/hr) to a flame ionization detector to measure total hydrocarbons. Following this measurement, another similar sample volume is admitted into a stripper column, which removes the relatively heavy nonmethane hydrocarbons and water. The effluent from this column, consisting of methane and CO, then enters a gas chromatograph for separation. The methane, which exits first, passes unchanged through a catalytic reduction tube and into the detector, where it is recognized as methane. Carbon monoxide, which exits next, passes through the platinum–hydrogen reducing atmosphere, and emerges as methane. It is thus detectable by the ionizing flame, where it is electronically recognized as CO.

Nonmethane levels for these sequential samples result from subtracting the signal of the methane hydrocarbons from the total hydrocarbons where nonmethane HC = total HC − methane HC.

Instruments equipped with various ranges for hydrocarbon detection are available. Ordinarily, the following ranges are preferable:

1. Total hydrocarbons—0 to 20 ppm.
2. Methane—0 to 10 ppm.

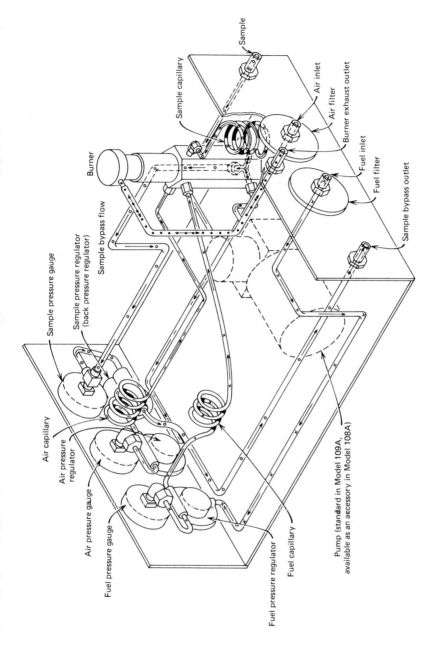

Figure 6. Flow diagram for total hydrocarbon analyzer. (Courtesy of Beckman Instruments Inc., Fullerton, Calif.)

These instruments can be made sensitive to as low as 0.1 ppm THC (as methane) and 0.05 ppm methane (CH_4).

Interferences

No interferences have been observed from any compounds that fall under the category of ionizable hydrocarbons.

Calibration

Calibration of these instruments is generally accomplished according to their operation as a "total hydrocarbon analyzer," with the necessary addition of CO, at a suggested concentration of 20 ppm, to establish the span setting. It is also a recommended practice to introduce span gases of intermediate concentration (i.e., 5, 10, and 15 ppm methane), and CO to verify the linearity of instrument response with an actual contaminant level.

Such a commercial instrument, as described, is illustrated schematically in Figure 7.

Suggestions in the operation of these instruments include repacking or replacing of the stripper column after approximately 2 months of continuous use at 10 to 40 ppm, or the equivalent ppm-hour running time. Temperature of the instrument during the operation should be maintained at $70 \pm 3°F$.

Continuous Particle Monitoring

Since the most harmful respirable portion of atmospheric particulate is found in the size region of suspended particles below 1 μ in diameter and since these particles cause an effective decrease in visibility, recent efforts have resulted in several continuous aerosol monitoring instruments whose principle of operation depends on the photometry of particle physics.

Integrating Nephelometer

Light scattered at approximately right angles to its direction of propagation in air indicates physical interaction of the light waves with particles in their path. The intensity of this scattered light, collected across a critical scattering sector, is proportional to the number of particles in the path of the light beam. In the operation of this instrument, sample air containing suspended particles is pumped into a sensing chamber at a rate of approximately 0.1 ft^3/min. In some instruments, the sample air is mixed with filtered purge or "carrier" air, which aligns the sample air in the light path and aids in establishing laminar sample flow that is free of backup eddies.

The optical system through which the sample air passes is illuminated by a high-intensity visible light. Since the individual particles scatter light of ap-

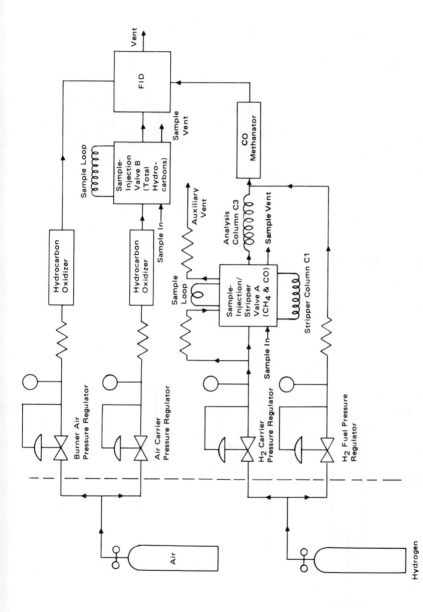

Figure 7. Nonmethane hydrocarbon analyzer flow and output diagram. Readout includes total hydrocarbons, methane, and CO. (Courtesy of Beckman Instruments Inc., Fullerton, Calif.)

proximately the same wavelength as their diameter, greater selectivity is achieved by using selected wavelengths to scan the sample to determine the particle population over several ranges of particle diameter. Simpler instruments employ a minimum of one or two wave settings, while more elaborate and more costly models permit as many as 10 to 20 channels or ranges in which to count particle response from particles as small as 0.1 to 03 μ in diameter to those in the 5 to 10 μ range.

Scattered light is collected by either parabolic or elliptical mirrors and focused to a photomultiplier tube detector, where the output of the detector tube is proportional to the population of particles scattering the light. Transmitted light that passes through the chamber without scattering is either reflected away from the container or absorbed by an optically inert cone. Calibration of the detector is achieved either by means of a fiber optic path that mechanically simulates scattering from a given population of particles (see Figure 8) or by actually admitting a measured population of known fixed particle sizes. These calibration standards, usually styrene grains, are commercially available.

Aerosol Vapor Condensation Meter

When very small particles, much smaller than 1 μ is size, enter a volume of air that is maintained at saturation humidity, fog traces are generated as the

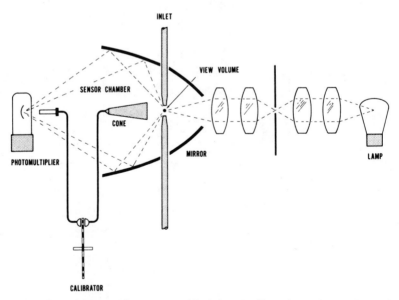

Figure 8. Optical system of continuous Nephelometer Particle Analyzer. (Courtesy of Climet Instruments Company, Redlands, Calif., Pat. U.S. 3248552, Can. 774850.)

particles nucleate tiny droplets of water vapor. If a small volume of sample air containing suspended particles 0.3 μ in diameter or smaller is admitted into such a humidity cloud chamber, the result is an instantaneous pattern of tracks representing the nucleation path of these particles.

Instruments based on this cloud chamber principle operate continuously by sequential preparation of the moisture-saturated chamber with the admission of sample and subsequent photometry of the haze tracks at intervals of 1 sec/sec. This rapid cycling is possible because of the almost instantaneous photometry involved.

A typical instrumental particle distribution range is 0.0025 to 1.3 μ diameter, with a population counting capability of 300 to 300,000 particles/CM^3.

Other Continuous Analyzers

Recent months have witnessed the introduction of continuous analyzers based on electrical resistivity, infrared emission, gas diffusion, and specific surface absorption of various contaminants in air.

Implied in the general tendency toward selection of physical measurements is the exploration of as many gas-phase physical properties as possible in an effort to arrive at a means of detecting and quantitatively measuring gaseous air contaminants.

Thus far, very few of these instruments have met with success in either accuracy or field reliability. Instruments based on the electrical signal that develops when specific absorption or diffusion of a gas-phase contaminant takes place on a membrane surface have suffered from the limitation imposed by deactivation of the surface. This disadvantage has been partially overcome by the introduction of "throwaway" membrane cells whose active life is more than 1 week and less than 2 months of average part per million per month exposure to particular contaminants.

In the future, it is hoped that continuous air analyzers for the halogens, hydrogen sulfide, odors, and many specific metals as particulates will be developed along such lines of approach. Until that time, it would not be fair to the reader to review specifically such continuous field monitors as might be wholly replaced by better and more reliable methods even as this text is written.

It is sufficient to say that as in any other analytical approach, the most exact methods for air analysis are those that are the most simple, involve the least number of steps or separations, and are most truly representative of the physical state and level of the air contaminant, in situ, as a gas-phase solution.

REFERENCES

1. M. D. Thomas and G. A. St. John, "Air Atmospheric Fluoride Recorder," *Am. Soc. Testing Mat., Spec. Tech. Pull.*, **250**, 49-57 (1958).
2. V. H. Regener, "Automatic Chemical Determination of Atmospheric Ozone," *Adv. Chem. Ser.*, **21**, 62-65 (1959).
3. P. O. Warner, S. Shaldenbrand, and L. Saad, Internal Report (unpublished) Wayne County Air Pollution Control Laboratory, 1971.
4. *Federal Register*, **36**, 8195, (April 30, 1971).
5. G. W. Nederbacht, A. Vander Horst, and J. Van Duijn, "Rapid Ozone Determination Near an Accelerator," *Nature*, **206**, 87-89 (1965).
6. M. B. Jacobs, M. M. Braverman, and S. Hocheiser, "Continuous Determination of Carbon Monoxide and Hydrocarbons in Air by a Modified Infrared Analyzer," *J. Air Poll. Control Assoc.*, **9**, 110-114 (1959).
7. R. G. Smith et al., "Tentative Method of Continuous Analysis for Carbon Monoxide Content of the Atmosphere (Nondispersive Infrared Method)," *Health Lab. Sci.*, **7**, 81 (1970).
8. Preliminary Summary, "Colloborative Testing of Reference Method for the Continuous Measurement of Carbon Monoxide in the Atmosphere (Nondispersive Infrared Spectrometry)," Southwest Research Institute for E.P.A., N.E.R.C. contact CPA., 70-40, January 7, 1972.
9. G. C. Ortman, "A Gas Chromatographic Approved to the SemiContinuous Monitoring of Atmospheric Carbon Monoxide and Methane," *Anal. Chem.*, **38**, 644-646 (1966).
10. A. B. Lamb, W. C. Bray, and J. C. Fraser, "The Removal of Carbon Monoxide from Air," *Ind. Eng. Chem.*, **12**, 213 (1920).
11. C. H. Lindsley and J. H. Yoe, "Simple Thermoelectric Apparatus for the Estimation of Carbon Monoxide in Air," *Anal. Chem.*, **2**, 127-132 (1948).
12. R. C. Robbins, K. M. Borg, and E. Robinson, "Carbon Monoxide in the Atmosphere," *J. Air Poll. Control Assoc.*, **18**, 106-110 (1968).
13. *Air Quality Criteria for Carbon Monoxide*, U.S. Department of HEW, P.H.S., NAPCA., Washington, D.C., Section 5, p. 5, March 1970.
14. M. D. Thomas et al., "An Automatic Apparatus for Determination of Small Concentrations of Sulfur Dioxide in Air," *Ind. Eng. Chem., Anal. Ed.*, **4**, 253-256 (1932).
15. M. D. Thomas et al., "Automatic Apparatus for Determination of Small Concentrations of Sulfur Dioxide in Air," *Ind. Eng. Chem., Anal. Ed.*, **18**, 383-387 (1946).
16. *Air Quality Criteria for Sulfur Oxides*, U.S. Dept. of Health, Education and Welfare, P.H.S., NAPCA., Washington, D.C., Chapter 2, p. 21, January 1969.
17. M. Terabe et al., "Relationships Between Sulfur Dioxide Concentration Determination by the West-Gaeke and Electroconductivity Methods," *J. Air Poll. Control Assoc.*, **17**, 673-675 (1967).

18. S. G. Booras and C. E. Zimmer, "A Comparison of Conductivity and West-Gaeke Analyses for Sulfur Dioxide," *J. Air Poll. Control Assoc.,* **18,** 612-615 (1968).
19. H. F. Palmer, C. E. Rodes, and C. J. Nelson, "Performance Characteristics of Instrumental Methods for Monitoring Sulfur Dioxide," *J. Air Poll. Control Assoc.,* **19,** 778-781 (1969).
20. R. K. Stevens, L. F. Ballard, and C. E. Decker, "Field Evaluation of Sulfur Dioxide Monitoring Instruments," Review, Research Triangle Park, N.C., 1972.
21. A. Fontijin, A. J. Sabadell, and R. J. Ronco, "Homogeneous Chemiluminescent Measurement of Nitric Oxide with Ozone," *Anal. Chem.,* **42,** 575-579 (1970).
22. D. L. Repley, J. M. Clingenpeel, and R. W. Hurn, "Continuous Determination of Nitrogen Oxides in Air and Exhaust Gases," *Air, Water Poll.,* **8,** 455-463 (1964).
23. G. M. Mast and H. E. Saunders, "Research and Development of Instrumentation of Ozone Testing," *Instr. Soc., Am. Trans.,* **1** (4), 325-328 (1962).
24. B. E. Saltzman et al., "Tentative Method for Continuous Monitoring of Atmospheric Oxidant with Amperometric Instruments," *Health Lab. Sci.,* **7** (1), 13, 22 (1970).
25. L. Potter and S. Duckworth, "Field Experience with the Mast Ozone Recorder," *J. Air Poll. Control Assoc.,* **15,** 207-209 (1965).
26. I. Cherniack and R. J. Bryan, "A Comparison Study of Various Types of Ozone and Oxidant Detectors Which Are Used for Atmospheric Air Sampling," *J. Air Poll. Control Assoc.,* **15,** 351-354 (1965).
27. D. H. Byers and B. E. Saltzman, "Determination of Ozone in Air by Neutral and Alkaline Iodide Procedures," *Am. Ind. Hyg. Assoc. J.,* **19** (3), 251-257 (1958).
28. *Federal Register,* **36** (84), 8195-8197 (1971).
29. V. H. Regener, "On a Sensitive Method for Recording of Atmospheric Ozone," *J. Geophys. Res.,* **65,** 3975-3977 (December 1960).
30. J. N. Pitts, W. A. Kummer, R. P. Sterr, and B. J. Finlayson, private research communication, 1971.

5

PRINCIPLES OF AIR SAMPLING

GENERAL CONSIDERATIONS AND APPROACHES

In the analysis of aqueous solutions for dissolved material two approaches are possible.

1. The analysis can be carried out in solution by aqueous color-forming, precipitation, or electrometric reaction.
2. The dissolved solute can be isolated by evaporating the water and then chemically treated selectively as a dry concentrate.

Air pollution usually presupposes a situation in which very small amounts of material are dissolved or suspended in air. Analysis then becomes a product of several physical approaches, like the analysis of aqueous solutions; these approaches attempt to determine the desired constituent by gas–gas or gas–liquid phase reaction, or by filtration.

In analysis of a conventional liquid, it is necessary only to transfer a viewable volume or weighable quantity of the liquid (aliquot) to a reaction vessel. In sampling gases (which exhibit no meniscus), the weight needed to activate a normal analytical balance at atmospheric pressure would very likely occupy the unwieldy volume of several liters. Consequently, air sampling, of necessity, requires a great deal of mechanical assistance to accurately measure large volumes of air from which a very small amount of contaminant is at some point extracted or determined through reaction.

GENERAL REQUIREMENTS OF A SAMPLING METHOD

Collection of a sample of pollutant in a quantity of air requires the following:

1. Use of an accurate flow device to measure the volume of air sample. Such a device must be calibrated as carefully as one would calibrate a piece of volumetric glassware, since these devices represent the measure by volumetry.
2. Use of a sample collector, generally a filter or an absorbing solution, to trap the contaminant material. The actual efficiency of the collector must be de-

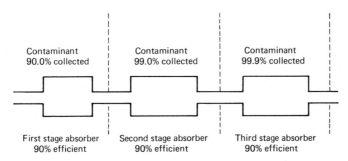

Figure 1. Increased efficiency of sample collection through the use of absorbers in series, each 90% efficient.

termined experimentally, so that the analyst can calculate the true weight or volume of contaminant as if all, or 100%, had been trapped. In this respect, very few sample collectors operate with 100% collection efficiency, although collectors arranged in series often approach perfectly efficient sample collection by combining their efficiencies, as shown in Figure 1.

Efficiency is best determined by using a known contaminant–air mixture as a control sample. The percentage of the contaminant recovered in the absorber (e.g., 80%) thus yields the working efficiency.

If no "known" contaminant is available, a rough relative efficiency can be determined if one assumes 100% collection in the last collector in a series of n collectors. Then the relative difference between any actual values for contaminant determined in $(n-1)$ collectors permits calculation of a relative efficiency, assuming no dilution effects or variation in the rate or mechanism of collection.

It should be noted that such sample collection problems as too small particle size or poor chemical reactivity may result in no collection in a backup sampler that suffers from the same inherent limitation as the first sampler, thereby erroneously suggesting 100% efficiency. One of the best sources of approximate sampling efficiency ranges is manufacturers' manuals supplied with filters, collectors, or collector assemblies.

3. Use of a pump that will deliver a constant flow of sample air to the collector. Marked variance in the rate of sample flow due to pump fatigue in the course of sampling will lead to erroneous sample volumes when time-averaging is used to calculate sample volume from the measured "initial" and "final" flow rates and the time of flow.

Ordinarily, sample volume is measured using:

$$\text{average flow rate} \times \text{time} = \text{volume}$$
$$10 \text{ liters/min} \times 30 \text{ min} = 300 \text{ liters}$$

HOW TO AIR SAMPLE

Given the general equipment just described, one should have some idea of which air contaminants are present either from general knowledge of the type of industry or from observation of physical damage. For example, foliage and plant damage may indicate ozone or SO_2 pollution, while discoloration of paint on buildings is often a sign of hydrogen sulfide. Persistent brown haze accompanied by general decrease in visibility may mean elevated levels of NO_2.

Each of the foregoing pollutants are gases and are most often sampled by bubbling ambient air through an aqueous solution that will react with and thus collect the suspected gas, if present, in the air sampled.

Particles, which are sometimes visible as haze, are always present in normal air. Here, the kind and quantity of such particles, usually collected by passing air through a filter, indicates the degree of air pollution.

How Long to Sample

Air must be passed through the collection device for a long enough time to obtain a detectable sample. Most air sampling procedures, such as those presented in this book, are designed to measure the usual range of

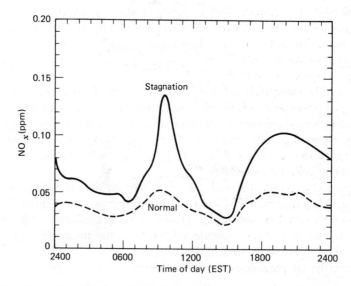

Figure 2. Diurnal variation of NO_2 levels during stagnation in Washington, D.C., October 19, 1963. (From reference 3.)

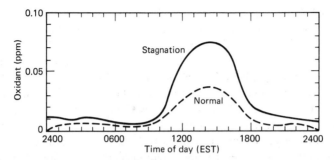

Figure 3. Diurnal variation of oxidant levels during stagnation in Washington, D.C., October 15 through 19, 1963. (From reference 3.)

contaminants found in "normally" polluted air. Thus a good general rule for contaminant gases is to trial sample for from 1 to 3 hr and then develop the color or otherwise analyze for the contaminant by the method specified.

Too large a sample, as shown by too dark a color on color development, can often be diluted and reread. On the other hand, too little sample usually requires a rerun of the collection procedure.

Ordinarily, samples collected at normal flow rates of 0.5 ft^3 of air per minute would result in the extraction of about 2½ m^3 of air in the suggested 1 to 3 hr. For the average contaminant, this would yield about 600 to 2500 μg of total material or approximately 30 to 125 μg of contaminant per milliliter of absorbing-collecting solution (allowing 10 to 100 ml of solution). This quantity is ordinarily sufficient for most colorimetric reactions of the kind generally employed in aerometry.

Short-Time Versus Extended-Time Sampling

Although governmental and industrial air analysts are usually interested in the "long-term" and overall effects of air pollution, the student, or perhaps, neophyte environmentalist, often seeks evidence confirming the existence of particularly high levels of pollutant, which signify fumigation conditions. It can be demonstrated statistically that the intensity and duration of these fumigation peaks in the so-called time-pollution curve have greater effects in determining the value of overall long-term averages than do a larger number of intervals showing moderately elevated pollution levels.[1] Such peaks often occur during periods of atmospheric stagnation, such as the examples shown in Figures 2 and 3.

As shown in Figure 2, several short sampling times of 1 hr each tend to produce average NO$_2$ levels quite different from the 24-hr average. It is therefore important to express the level of pollutant observed for precisely the time sampled. There are, of course, some exceptions to this rule, notably in cases where chemical stability of the sampling solution precludes a full hour's sampling although as hourly average may be reported arbitrarily.[2]

Short-term sampling times during periods of obvious fumigation can reveal valuable information, such as the real-time levels observed in Figures 2 and 3, if it is clear from the reading of the data exactly how much sampling time is involved. The danger implicit here is that a novice may carefully sample for 1 hr during extremely high levels of pollution and then publish a report in which the pollution levels are expressed without regard to time, leading the reader to the conclusion that these levels represent long-term or even annual averages.

As an illustration, consider a case where measurements of CO made during peak expressway traffic times of 4:00 to 8:00 P.M. are reported as "levels of carbon monoxide observed on expressways." Granted that these measured concentrations represent levels of human exposure, such levels of perhaps 150 to 200 ppm of CO are not sustained for long durations by motorists who enter and leave the area. In addition, because of traffic cycles, such short-term data do not accurately reflect the true average CO level to which even residents in the adjacent area may be exposed over a 24 hr period.

Significance of Sampling Time With Respect to Biological Effects of a Pollutant

Since the human body behaves very much like the pollutant sampler in its ability to absorb a contaminant through chemical reaction, it is important for us to ask the reaction time of the human body to various contaminants, assuming that like the sampler, the body exhibits relative degrees of absorbing efficiency. Here, much remains to be discovered in understanding the dose-effect relationship between short-term exposures of humans to elevated levels of a pollutant and longer-term continued exposure to much lower levels.

Response of the body to a given level of pollutant follows from the biological half-life of the pollutant (i.e., the time during which one-half of the amount present in the body at time t_0 is removed). It has been found that exposure to lower concentrations of an air contaminant allows the body or organism the opportunity to restore a state of repair. It is thought that, since the concentration of pollutant in ambient air varies from time to time, the concentration in the critical absorbing body organ adjusts exponentially, reaching equilibrium concentration after three or four half-lives.[3] This, of course, implies a factor of time duration on the toxicological effect of a pollutant.

Assuming that the rate of pollutant absorption by the body is proportional to its ambient concentration, Roach[4] has stated that variations in ambient pollutant concentrations over less time than one-tenth of the biological half-life have no biological consequence.

Table 1 Biological Half-Lives of Air Pollutants for Either the Whole Human Body or the Most Sensitive Organ[5]

Pollutant	Biological Half-Life
CO	2 hr
Chlorine gas	<20 min
H_2S	<20 min
Iron oxide particulate	12 hr
Lead	6 months
Mercury	5 weeks
NO_2	1 hr
SO_2	<20 min

We may therefore infer that the same relationship holds true for sampling time. It is not crucial for us to sample continuously, minute by minute, for a pollutant whose biological half-life is shorter than 10 min. The biological half-lives of several common air pollutants are given in Table 1.

Limitations of the Air Sampler

Probably the most important rule in sampling is to sample long enough and at a rate sufficient to collect an analytically measurable sample. We have previously discussed the ordinary duration of sampling and suggested a time interval of 1 to 3 hr to collect a minimum of 30 μg of sample per milliliter of solution—a weight that is normally sufficient for chemical measurements.

Although the optimum rate of sampling varies from collector to collector, it is possible to establish some general guidelines. Most sample collectors for gaseous air contaminants require an airflow of between 0.05 and 1.0 ft^3/min. A simple inexpensive diaphragm pump is able to sustain a flow rate of 0.05 to 0.10 cfm, while a modest-size rotary vacuum pump may be required to produce a dependable airflow of 1.0 cfm or greater. For particulate sampling, there is an even greater discrepancy between the limits of permissible sampling rates. The low-flow electrostatic collector requires only a fraction of a cubic foot of air per minute, while the high-volume air sample pumps as much as 50 to 60 cfm through a glass-fiber filter.

At this point, it might seem reasonable to assume that the greater the sampling rate, the more sample will be collected in a given sampling time. In this

Table 2 Flow Rates for Various Samplers

Contaminant Sampler	Optimum Flow Rate (cfm)
Gases and Vapors	
Midget impinger	0.05–0.10
Standard impinger	0.5 –1.0
Particulate	
Membrane filter	0.5 –4.5
High-Volume filter	40–60

respect, it is important for the air chemist to realize that the maximum flow rate is not often the optimum flow rate. Too fast a flow through an air filter may well rupture the filter. Elevated flow through a water-filled collector (impinger) may well result in channeling of the air through the collector, with subsequent loss in efficiency proportional to the loss in effective liquid surface area.

The average optimum flows for the four most common sampling devices are as shown in Table 2.

In using these suggested flow rates, one must avoid the assumption that the collection efficiency of a given collector remains the same no matter what material is being sampled. Indeed, the efficiency of a midget impinger for aerosol dust at 0.06 cfm may be quite different from that for gases at the same flow rate. Ordinarily, bubblers or gas samplers must operate above a given threshold flow rate of about 0.01 cfm. Below this rate, collection efficiency is greatly decreased because of lack of surface contact; too little sample air forms large bubbles that permit little or no contact of the body of the bubble, which contains sample air, with the absorbing liquid.

Collecting the Contaminant as It Exists in Air

It should be born in mind that contaminants collected by normal air sampling methods may or may not remain unchanged after collection. For example, air drawn across already collected particulate on the face of a filter is capable of oxidizing the entrapped mineral or organic compounds in the collected contaminant dust. Active aerosols, such as sulfites, are subject to weight change as a result of reactions with water vapor (humidity) and/or air oxygen in the presence of other mineral catalysts, such as alumina or silica, that are normally coentrained in the filter medium. Also, collection of SO_2 by bubbling through water without prior separation of ambient particulate can result in oxidation of the SO_2 to SO_3 by air oxygen due to the catalytic action of mineral particulate.

Consequently, when it is necessary to collect and identify specific chemical compounds, precautions must be taken to maintain the chemical integrity of the desired compound. For example, it may be necessary to use a reducing agent, such as isopropanol, as a medium for collecting SO_2 to inhibit the formation of SO_3 by the reaction:

$$2SO_2 + O_2 \rightarrow 2SO_3$$

Even the collection vessel is capable of surface reaction with collected sample. Probably the most common such interaction is the adsorption of gases on the walls of the collecting container, such as a plastic bag or metal probe. For this reason, air analysts, usually choose inert nonabsorbant polyfluorocarbons (e.g., Tedlar) over polyolefin or vinyl materials. These latter polymers possess more polar double-bond systems, which may collect or chemisorb such gases as SO_2 or NO_2.

Prolonged storage of gas or liquid samples is discouraged because of any concentration changes that may result from reaction of the desired constituent gas with other gases in the mixtures as well as with the walls of the container. When such storage is necessary, either glass or Teflon should be chosen as the material for storage containers, with preferable storage conditions being refrigeration and darkness.

Assembling Materials for Air Sampling

Except for the continuous analyzers described in Chapter 4, most air sampling is performed using laboratory glassware. Such sample collecting is often performed with both the operator and the majority of the glassware inside a building or shelter. A glass or polyfluorocarbon probe is used to connect the sampling apparatus to the outside from which sample air is drawn. The system usually consists of the following:

1. The sample collecting device, such as an aqueous solution-filled cylinder containing a tube that reaches to the bottom, or in the case of particulates, a 0.1 to 10.0-μ pore size glass-fiber filter on a metal support screen.
2. A device to measure airflow through the collector (or absorber).
3. A pump to draw sample air through the system.

A typical arrangement of these components is shown in Figure 4. It should be noted that the pump is always placed so as to draw, rather than push, air through the collecting device. This is mainly to prevent the possible contamination of collected sample with pump lubricant or vane material. The airflow device may be connected electronically to a recorder so that variations in flow due to such variables as fluctuation in power line voltage can be recorded and accounted for in assessing the total volume of air sampled.

Principles of Air Sampling

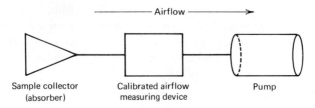

Figure 4. Arrangement of simple air sampling equipment.

Common Units of Measurement Employed

Standardized test procedures very often contain definitions of airflow and weight of material that require the use of specific units of measurement to express air sampling results. Examples of such conventional units are shown in Table 3.

Selection of Sampling Site

After consideration has been given to the sampling time, sampling rate, and sampling procedure, an actual selection of the sampling site must be made.

The human nose, which is the main channel through which air pollutants enter the body, normally operates at a distance of between 5 and 6 ft from the ground. Therefore, it would seem appropriate to sample air at an elevation within this range. In the case of gases or vapors, this may well be an optimum sample probe height. However, when a sampler is to remain on site and

Table 3 Commonly Used Units of Air Sampling Measurements

Measurement	Units
Suspended particulate	Micrograms per cubic meter of sample air
Gases	Micrograms per cubic meter of sample air or parts per million by volume
Metal particulate	Micrograms per cubic meter of sample air
Settleable particulate	Pounds per thousand square feet per 30-day period or tons per square mile per 30-day perild
Temperature	Degrees centigrade
Pressure	Millimeters of mercury
Sample airflow	Liters per minute or cubic feet per minute

operate intermittently, as is frequently the case, a better arrangement is often achieved by placing the sampler above ground level in a building at a suitable location near a window through which a sample probe (Teflon tubing) can be positioned at 10 to 15 ft above ground level. This arrangement also discourages the physical tampering with the probe outlet by a child or curious passerby. The general arrangement of such sampling equipment is as shown in Figure 4.

AIR SAMPLING EQUIPMENT

We have discussed the general arrangement of the three main pieces of air sampling equipment—air pump, airflow measuring device, and sample collector (or absorber). We now examine in some detail the properties of these pieces of equipment and their operating characteristics.

The Pump

The air pump should always be the last unit in a "train" or series of air sampling components. It can be powered by line voltage or electric current from a battery converter. Commercially available pumps vary considerably in their flow capacity, reliability, and principle of operation. A simple division of pumps can be made on the basis of the performance required.

Diaphragm Pumps

Gas sampling using small "midget" absorbers that contain only 10 to 20 ml of absorbing solution require a rather low flow (1 liter/min) of sample air and consequently employ a rather simple pump. Such units are called diaphragm pumps, in which the diaphragms are made of metal or a synthetic composite. These pumps allow as much as 2 to 3 liters/min (up to 0.1 cfm) of sample airflow. They are commonly employed in intermittent, routine, on-site, air sampling. Diaphragm pumps can be operated using either line voltage, an automotive-type lead storage battery (with a power converter), or the power generator system of an automobile or small airplane. (In sampling close to an internal combustion engine of any kind, caution should be exercised to prevent inadvertent sampling of exhaust fumes.)

Rotary Pumps

For rapid sampling (>1 cfm), where very low ambient concentration of a contaminant or the need for short sampling time requires gathering a large sample volume in a short time, a rotary pump is required. These pumps may be either oil lubricated or contain graphite vanes that employ actual wear of the vanes against the cylinder walls to provide essential lubrication. Vanes are thus consumable without loss in pressure, but must be replaced occasionally because of excessive wear or fracture. These pumps are capable of providing 20 to 30 liters/min (about 1 cfm) airflow or producing a pressure drop of up to 140 to 160 mm Hg across a limiting orifice (see Chapter 3, Determination of Nitrogen Dioxide by the Jacobs-Hochheiser Method). Failure of such a pump to maintain within 10 to 15% of its original or design pressure indicates a possible need for replacement of one or more vanes.

Airflow Measuring Devices

As mentioned earlier, when ambient air is subject to analysis, accurate and reliable measurement of the volume of air sampled is of equal importance with the chemical determination of the desired contaminant. This measurement is best carried out either upstream from or immediately following the actual sampling device. Flow measurement downstream from the pump should be made under the most unusual circumstances (see Figure 5). Measurement of the airflow downstream from the pump is subject to error due to the following:

1. Deposition of vane material or pump lubricant into the air-measuring device.
2. Inclusion of pressure drop inherent in the pump in the airflow data.
3. Temperature change in the air stream, which causes a change in the volume of air measured.

Simple Direct Methods of Air Volume Measurement

Measurement of Sample Volume by Single Displacement

A "spirometer" uses the principle of volume displacement to measure the exact amount of air passed through sampling equipment by measuring the volume of water displaced by such air at a known temperature. Although this is a very accurate measurement, it involves prolonged contact time of the air sample with the walls of a large container and requires a very large measured volume (10 liters). Its size and initial cost make this device impractical for routine air measurements. However, it is valuable as a means of standardizing other air pollution equipment.

A simple spirometer (a displacement bottle) can be prepared from a jug or cylinder whose exact volume at a given temperature is known or calibrated by the weight of water contained.

The displacement method is primarily used to standardize portable field air metering equipment (Figure 6). It should be reemphasized that the spirometer as well as other pieces of primary standard volume measuring "equipment"

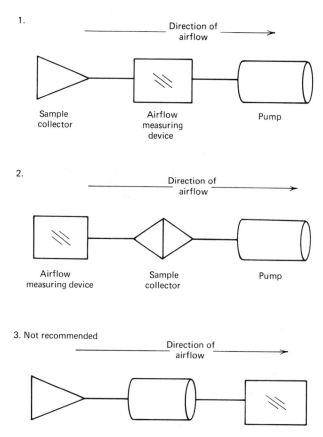

Figure 5. Positioning of the airflow measuring device.

discussed here are most appropriate for in-laboratory and primary calibration rather than for actual field calibration.

Measurement of Sample Volume by Continuous Displacement

THE DRY GAS METER. Two bellows that are alternately filled and emptied provide the basis for actuation of a set of dials that display the volume of air passed, in liters or cubic feet. A stopwatch is used to time the duration in which a given volume, say 20 liters, passes in a total time of 10 min to yield a flow of 2 liters/min.

These instruments are ordinarily available in capacities appropriate for calibration of airflows from 0.3 to 30 cfm (see Figure 7).

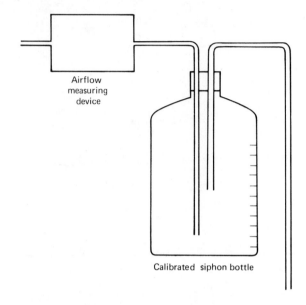

Figure 6. Use of a siphon bottle or simple spirometer to standardize the exact volume of air passed through airflow measuring device.

THE WET TEST METER. Somewhat more accurate than the dry gas meter, but not as accurate as the spirometer, the wet test meter is capable of measuring airflow to an accuracy of $\pm 0.5\%$ of the total volume.

As shown in Figure 8, the wet test meter consists of a series of chambers mounted radially around a central axle, and most of the chambers totally immersed in water, which serves to seal the rotating chambers for an accurate measure of water displacement. Entrapped air passes through the water-filled section as the axle rotates and is released above the surface of the water through a valve at the top of the assembly. The rotating axle thus actuates an index dial that displays air volume as a function of the number of revolutions of air-filled chambers of known volume.

As in the case of the dry gas meter, rate of airflow is obtained by timing with a stopwatch the minutes required for a given volume of air to pass through the meter.

Indirect Methods of Air Volume Measurement

"Rate-of-flow" measuring instruments are ordinarily compact and quite suitable for field use. However, they must be calibrated using a stopwatch together with either the wet test meter or dry gas meter. Such devices are usually placed in the sampling system for just long enough to momentarily measure the flow rate. They are then removed, leaving only the pump and

Figure 7. Dry gas meter.

Figure 8. Wet test meters (Courtesy of Precision Scientific Co., Chicago Ill.)

210   Principles of Air Sampling

sample collector to operate as the air sampling train. Thus one must remember that such an intermittent measurement of a system's flow rate does not ensure accurate measurement of the total sample volume. Interruption of sampling time due to voltage loss or temporary power failure would remain undiscovered if only initial and final flow rates were determined and averaged to calculate average flow. This would, of course, lead to error in calculating the final volume of sample air from relationship:

$$\frac{\text{liters}}{\text{minute}} \times \text{minutes} = \text{liters}$$

$$\text{average rate-of-flow measurement} \times \frac{\text{sampling}}{\text{time}} = \frac{\text{sample}}{\text{volume}}$$

This disadvantage in the use of rate-of-flow equipment can be minimized by adding a voltage-time recorder to the electrical circuit that powers the pump.

Other factors that may similarly influence intermittent measurements made using the flow-rate device are plugging of the air sampler or change of ambient air temperature or pressure. These detrimental effects can be avoided only through the use of a continuous rate measurement device with attached time-flow recorder.

Simple Flow-Rate Meters

MANOMETER. The differential static pressure as a result of airflow past an orifice can be measured using a manometer.

Arranged as shown in Figure 9, the manometer itself is first calibrated through measurement of a known flow by the placement in series with a calibrated dry gas meter, followed by the intended air sampling device in the sampling train together with the sample air pump. A plot is then made of actual measured flow rate in liters per minute (dry gas meter) versus the ob-

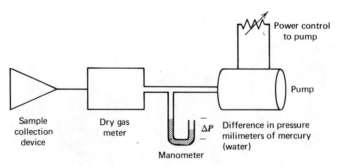

Figure 9. Use of manometer to measure flow by pressure differential.

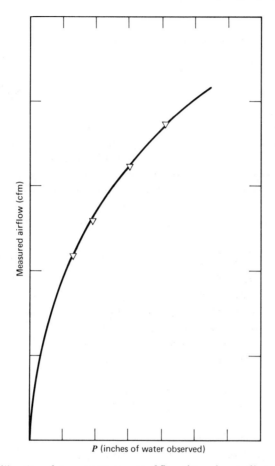

Figure 10. Calibration of manometer to rate of flow through sampling systems.

served difference in mercury level (manometer). Such a plot is shown in Figure 10.

The power control shown in Figure 9 is used to regulate the flow through the pump if a range of pump speeds is desirable. If only one pump speed setting is possible, only one point will appear on the graph (i.e., the actual pump speed) that will produce a measured flow (liters per minute by dry gas meter) at the observed manometer reading. Since the sampling system is subject to loss in flow rate because of filter plugging or constriction because of adsorption in the sample collector, manometer readings for various in-system pressures must be obtained using the calibration system described in the section Calibration of Flow-Rate Meters.

212   Principles of Air Sampling

Although manometer readings may be used to calculate flow directly, in-system calibration is generally the recommended approach to ensure reliable field measurements.

ORIFICE METER. An orifice meter is a pipe into which has been placed a metal disk having one (or more) holes (critical orifice) so that air passing through the disk establishes a measurable pressure drop across the disk that is proportional to the flow rate. As shown in Figure 11, pressure taps before and after the disk are used to measure the magnitude of the pressure drop. A simple stylus or dial can be actuated by this differential, or, as shown in Figure 11, a manometer can be used to establish the magnitude of the pressure drop.

The orifice meter is accurate to about 2% of the flow measured. It is subject to error due to accumulation of airborne soil or hydrocarbons in the orifice holes. Such constriction and plugging is indicated by slowly decreasing flow-rate readings with time.

ROTAMETER. When a lightweight float ball is placed in a vertical air stream in a graduated, slightly tapered glass tube, the ball will rise to a height determined by the rate of airflow through the graduated tube (see Figure 12). The spinning ball is carried upward by the current of air until it reaches a point where its own weight is balanced by the supporting force of the airflow. If this airflow is measured in series with the sample collector, as shown in Figure 13, and the rotameter is calibrated versus a dry gas meter in series with this flow, the rotameter is then calibrated to the sample system and can be used as a field device to check the sample airflow of the sampling system once a plot has been established for various rotameter readings versus actual flow rate through the system.

The rotameter is often placed ahead of the sample collector (Figure 13a) when an extremely long sample probe is necessary for access to the outside air. The rotameter is then subject to the pressure drop due to viscosity caused by air

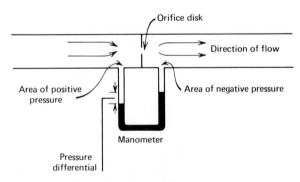

Figure 11.   Flow through an orifice meter.

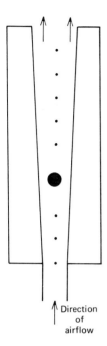

Figure 12. Flow of air through a rotameter causes the ball to float to a height dependent on the rate of flow.

turbulence in the probe. Alternatively, the rotameter is placed ahead of the pump following the calibration device (dry gas meter), as shown in Figure 13b.

Rotameters are generally accurate to 2% of the measured flow and are most subject to error from entrained particulate soil and hydrocarbons that deposit both on the walls of the vertical tube and the surface of the ball. Most rotameters incorporate glass or abrasion-resistant clear plastic flow tubes. The tube should be cleaned periodically with methanol when the ball exhibits sticking to the walls of the tube.

PITOT TUBE. The Pitot tube is based on the relationship between flow rate and dynamic versus static pressure in a system. (Figure 14)

Actual use of the Pitot tube is limited to rather high velocities (i.e., above 800 to 1000 fpm). Ordinarily, differential pressure is measured by means of an inclined manometer or a calibrated orifice meter. The Pitot tube lends itself best to measurement of high in-stack or in-duct flows rather than the sample flows normally encountered in air sampling. It is mentioned here, however, because of its application as a means of determining the velocities in ducts and stacks containing air to be sampled. Here, the measured velocity in linear feet per minute is multiplied by the area of the stack to obtain the emission rate to the atmosphere of sample air in cubic feet per minute. Thus an in-stack air

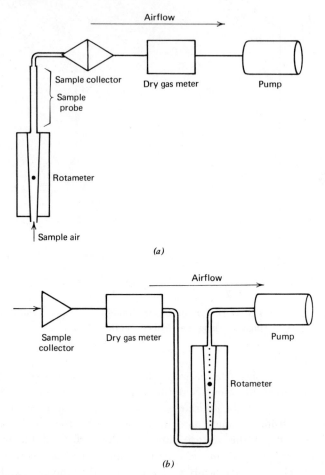

Figure 13. Two arrangements of a rotameter.

sample that is found to contain SO_2 at a concentration expressed in micrograms per cubic meter, may be multiplied by the flow rate converted to cubic meters per minute to yield the source strength in micrograms per unit time.

Calibration of Flow-Rate Meters

Airflow measurement devices are subject to error as a result of exposure to:

1. Sudden change in temperature.
2. Corrosive sample air.
3. Highly soiled sample air.
4. General misuse.

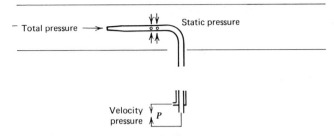

Figure 14. Pitot tube inserted in a duct or stack.

Therefore, frequent calibration is mandatory. For routine use (2 to 3 times per week), a piece of airflow equipment will probably require calibration at least three to four times per year. Here, a simple water displacement approach similar to the siphon bottle (described in Figure 6) is used by ASTM as a calibration method (ASTM D 1071-55).[6]

Equipment most often on hand in an air pollution laboratory includes a wet test meter (primary standard), which is used to calibrate a dry gas meter. The dry gas meter can then be used in series with a flow-rate field meter to calibrate the field meter on the laboratory bench before the equipment is carried into the field for actual air sampling. A dry gas meter is calibrated in the following manner, using the arrangement of equipment in Figure 15 as a guide.

PROCEDURE

1. Adjust screw clamp valve to open and, with equipment arranged as in Figure 15, start pump. This will at first permit the least amount of air flow through the 2 meters. Record the pressure and temperature of each meter as shown on the meter dials.

2. Using a stopwatch to determine time and the dial on the face of each meter to record flow, determine the time required for 1 ft^3 of air to pass through each meter, and tabulate trial 1 as shown in Table 4.

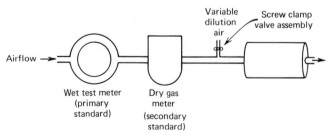

Figure 15. Calibration of dry gas meter.

Table 4 Example of Dry Gas Meter Calibration Chart

Trial	Wet Test Meter Rate (cfm)	Dry Gas Meter Rate (cfm)
1	0.006	0.004
2	0.014	0.009
3	0.060	0.040
4	0.150	0.100
5	0.281	0.180
6	0.660	0.441

3. Tighten the screw clamp successively to allow a greater portion of the airflow through both metering devices. Redetermine the flow rates of each trial and tabulate as in Table 4. Repeat this step at least four or five times until flows approach 1 cfm, maximum flow through most assembled air sampling equipment.

4. Prepare a calibration curve by dividing each wet test meter reading by the corresponding dry gas meter reading to determine a correction coefficient. Plot the observed reading versus actual corrected flow rate as shown on the wet test meter. If the same coefficient slope is obtained for all points within the range of flows covered, this coefficient can be used to correct observed dry gas meter data within this range when the device is used to calibrate the rate of field flow meters.

When possible, the entire field sampling train containing the field sampling meter should be calibrated in the laboratory, using the primary standard meter ahead of the field sampling system so that all individual instrument variables in this sampling unit will be accounted for.

The Air Sample Absorber

Actual collecting devices, as a class, act in some way to retain the desired pollutant either alone or as a mixture. Most of these devices merely collect the pollutant, and analysis must be performed on the collected sample at a later time.

Selecting of the proper sampler for a particular pollutant depends partly on the pollutant and partly on the equipment at hand. Individual sample collectors are generally divided into those that can be used to collect gaseous pollutants or particulate pollutants.

SAMPLING GASEOUS AIR POLLUTANTS

The bases for selecting a collector are discussed in the following sections, Sampling Gaseous Air Pollutants and Sampling Particulate Air Pollutants.

SAMPLING GASEOUS AIR POLLUTANTS

Although the most common gaseous pollutants are SO_2, NO_2, SO_3, O_3, H_2S, HF, and HCl, other vaporous material (e.g., condensable hydrocarbons and chlorinated hydrocarbons together with various functional group organic compounds) are usually classified as gaseous from the point of view of sampling procedures, with the only distinction being that vapors (which by definition exist below the critical temperature) are readily collectible by merely cooling the gas stream.

In practice, however, principles involved in collection of pollutant gases require a collector that will:

1. Absorb the gas by chemical reaction.
2. Adsorb the gas on a finely divided surface.
3. Convert the gas to a liquid or solid by cooling.
4. Capture a quantity of air (in a container) from which the pollutant gas will be extracted at a later time by one of the methods above.

Absorbing the Pollutant Gas in Reagent Solutions

Solution absorption, probably the most frequently employed method to collect gaseous pollutants, separates the desired gas from air by capturing it in a liquid either through direct solubility or by chemical reaction. The collector hardware is designed to increase the surface area of the liquid either by bubbling or repetitive recycling for the most efficient collection of the gas.

Factors that affect gaseous sample collection by absorption are:

1. Size of bubbles formed in absorber. Small bubbles yield greater surface area and allow greater contact across the air-liquid absorbing interface. Fine frits are therefore desirable in cases where high collection efficiency is required.

2. Contact time. Absorbers of large volume permit a longer contact of the pollutant gas mixture with the liquid surface of the absorbing medium. Lower flow rate of sample air also favors longer contact time, but the flow rate must be high enough to permit proper agitation of the liquid in the collector.

3. Concentration of the absorbing solution. Since the law of mass action predicts greater degree of reactivity with increasing concentration of the reactants, greater collection efficiency is generally observed with increasing concentration of chemically reactive absorbant.

4. *Speed of chemical reaction.* When a chemical absorption reaction is used to fix the pollutant in the collecting solution, the inherent speed of the reaction contributes to the collection efficiency. An ideal chemical collecting system would employ the most irreversible chemical reaction, such as the formation of a very insoluble precipitate. For example,

$$SO_3 + H_2O \rightarrow H_2SO_4$$
$$H_2SO_4 + BaCl_2 \rightarrow \underset{\text{(insoluble)}}{BaSO_4} + 2HCl$$

or an extremely stable complex ion

$$CO + Cu_2Cl_2 \rightarrow Cu_2Cl_2 \cdot 2CO$$

Choosing the Absorbing Liquid

Since water is the most commonly available absorbing medium, it is usually advantageous to capitalize on the relatively low solubility of air in water and merely pass air to be sampled through a given volume of water by bubbling. Table 5 compares the solubility of several pollutant gases in water to the solubility of an equal volume of air in water.

Obviously, such relative solubilities represent optimum collection conditions, since system equilibrium is implied. Although distilled water is suitable as a collection solvent for the acidic gases such as SO_2, the weaker acid gases (e.g., NO or CO_2) are best collected in 0.01 to 1.0 M sodium hydroxide solutions. Here, caution must be exercised in cases in which oxygen in the air may cause oxidation of the collected acidic anhydride in basic solution. For example,

$$2SO_2 + O_2 \xrightarrow{\text{basic solution}} 2SO_3$$

Here, if SO_2 is known to be free of SO_3, the resulting absorbing solution is merely analyzed for sulfate after addition of bromine water to ensure the complete oxidation of SO_2. If, however, the SO_2–SO_3 mixture is to be collected,

Table 5 Relative Aqueous Solubilities of Pollutant Gases

Pollutant Gas	Volume Solubility in Water Compared to an Equal Volume of Air in Water at 20°C
SO_2	876.
CO_2	19.1
H_2S	55.8
NO	1.02

then 80% isopropanol is used to collect SO_3 in the first impinger, while at least one backup impinger containing 3% hydrogen peroxide is employed to absorb SO_2, which is insoluble in the isopropanol.

In general, the best collecting situation results from a choice of absorbing solution that will react irreversibly with the given pollutant gas, so that no vapor pressure of the gas will exist in the aqueous system. Thus, CO_2 reacts irreversibly with lime water to form calcium carbonate.

$$CO_2 + H_2O \rightleftarrows H_2CO_3$$

$$H_2CO_3 + Ca(OH)_2 \to \underset{\text{(insoluble)}}{CaCO_3} + 2H_2O$$

As a rule, it is desirable to employ an excess of reactant [e.g., $Ca(OH)_2$ in the collector as high as 5 to 10% to ensure the best possible collection efficiency, especially in cases where prolonged sampling ($>$ 1 to 2 hr) may produce loss of absorbing strength as the reaction ensues].

Efficiency of the Collector

Methods exhibiting collection efficiencies of less than 50% are generally less desirable as quantitative approaches. This is due to the fact that lower actual observed sampling efficiency magnifies errors in the reproducibility of sampled gas volumes. Thus a 1% absolute error in the gas volume at 100% efficiency becomes a 3% final error if the collection efficiency is only 33.3%.

Determination of collection efficiency is accomplished by the procedures described in the section General Requirements of a Sampling Method in this chapter. The recommended procedure for determining the collection efficiency of a given wet (solution) absorber employs a stream of the desired gas in known concentration with no interference, preferably using clean dry air or dry nitrogen as the carrier gas.

Procedure for Determining Collection Efficiency

1. Arrange the sample collector (impinger) in exactly the manner to be used for field collection.

2. Using an accurate flow-measuring device and suitable pump, allow a flow of the calibration standard gas through the collector for exactly the time to be used in the field (expected low pollutant levels, 3 to 24 hours; expected moderate to high levels, 0.25 to 3 hours).

3. Using the sampling time and known concentration of pollutant gas ($\mu g/m^3$ or ppm), calculate the exact amount of pollutant to pass through the collector. For example, for a standard containing 10 $\mu g/m^3$ of SO_2 at a sampling time of 10 min using a measured flow rate of 0.1 m^3/min.

$$0.1 \, \frac{m^3}{min} \times 10 \text{ min} = 1 \text{ m}^3 \text{ total sample volume or } 10 \, \mu g \text{ of } SO_2$$

220   Principles of Air Sampling

If subsequent analysis of the absorbing solution should show 8.5 µg of SO_3, the collector, as assembled, exhibits 85% collection efficiency.

Comparison of Various Absorption Devices

In general, the absorption devices referred to in previous sections are composed of:

1. A collecting tube or bottle (usually glass).
2. A liquid-absorbing medium.
3. A "dispersion tube" or bubbler.

These collecting tubes fall into two general classes called fritted bubblers and impingers. Typical flow rates through these devices carry from 0.1 to 1.0 ft^3/min. Although available commercially, simple absorbers can be prepared in the laboratory from common glassware using a test tube or Erlenmeyer flask, as shown in Figure 16.

The shape of the flat-bottomed flask makes its use quite suitable as an impinger. The test tube, although practical, permits undesirable channeling of the airflow up the walls of the tube rather than acting to break the airflow into finer bubbles against the bottom. Consequently, the use of a test tube impinger is recommended only for cases where qualitative evidence of the presence of a pollutant is sought.

Quantitation of the laboratory flask absorber may be achieved using the simple siphon or water-displacement bottle, shown in Figure 6, to determine the rate of sample airflow through the absorber over a fixed period of time. In

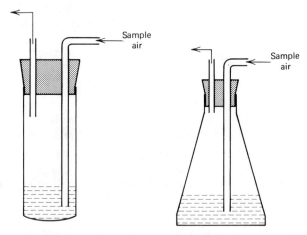

Figure 16.   Test tube, left, and Erlenmeyer flask, right, are used to fashion a simple absorber.

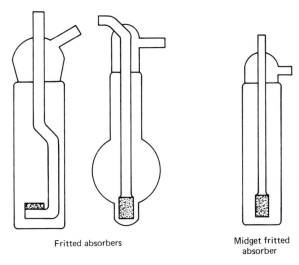

Fritted absorbers   Midget fritted absorber

Figure 17. Fritted glass absorbers.

general, a flow rate of no more than 0.5 cfm should be employed to collect an air sample in the Erlenmeyer flask.

Fritted-Glass Absorbers

In general, porous glass provides the greatest efficiency for gas sample collection. However, because of chemical surface reactivity, a fritted-glass bubbler is not usually employed to collect strong oxidizing gases such as ozone. Frits designated "coarse" (50-μ pore size) are most often used for air sampling. Figure 17 shows several varieties of fritted glass absorbers.

Impingers

The mechanics of breaking an airstream into bubbles requires turbulence. Such impact turbulence can be achieved by allowing the airstream to strike a flat surface a short distance from its path of exit in the absorbing device. Impingers, shown in Figure 18, make use of this principle to collect gases with efficiencies often approaching those of fritted-glass bubblers. An advantage of the impinger is the ease of cleaning the sampling equipment. Most impingers, as well as bubblers, operate with absorption efficiencies in excess of 75%; under optimum flow conditions, these efficiencies approach 90 to 100%.

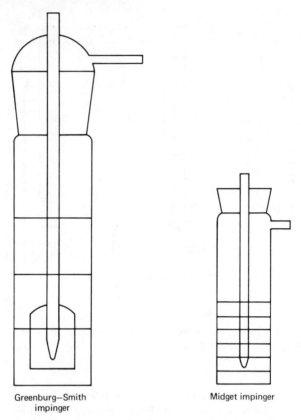

Greenburg–Smith impinger Midget impinger

Figure 18. Common impingers.

Adsorbing the Pollutant Gas on Finely Divided Solid

At moderate (30°C) to low (< 0°C) temperatures, gases adsorb on solid surfaces to a degree approximately inversely proportional to their boiling points. The tendency of a gas to so adsorb increases with the surrounding pressure and decreases with increasing temperature. However, adsorption behavior is most often represented as a function of pressure alone, keeping the temperature constant. Such graphic representations are thus called isotherms.

Ordinarily, gas adsorption proceeds rapidly at first exposure. Then adsorption proceeds at a slower rate as the adsorbant's surface becomes coated with gas molecules.

For the reader who wishes to pursue the molecular kinetics of these adsorption processes, the Freundlich and Langmuir relationships[7] provide good theoretical starting points.

Aside from the more theoretical considerations, other practical factors that must be weighed by the air analyst in judging the general suitability of physical adsorption as a collection method are:

1. The amount of adsorbant material. Greater capacity requires increased amount of total adsorbing substrate.
2. Mechanical state of the adsorbant. A large total surface requires very small particle size.
3. Overall efficiency. The greater the affinity of the pollutant gas for the substrate, the greater the adsorption efficiency.
4. Temperature. Low temperature generally favors adsorption.

A number of solids have been used successfully as adsorbing substrates, and with the advent of gas chromatography, various combinations of dry adsorbing materials have also been the subject of experiment to optimize adsorbant properties. In general, the most commonly used adsorbing materials are charcoal, silica gel, and alumina. These are activated by heating at $>110°C$ for a period of several hours to several days to increase their adsorbing activity. For charcoal especially, care must be taken not to overheat and thus to cause possible oxidation of the carbon.

In adsorption sampling, sample air is passed through a packed column or "canister" containing the finely divided solid adsorbant. If necessary, the column can be maintained below room temperature by means of cooling coils or a dry ice bath. Such an arrangement is intended to allow the generally higher-boiling pollutant gases (e.g., H_2S, SO_2, hydrocarbons) to adsorb, while the lower-boiling fractions of air (N_2, O_2) pass through the medium unchanged. At the end of a 1 to 6 hr sampling time, a mixture of pollutant gases is usually fixed on the solid adsorbant. These must then be removed by heating ($>100°C$), vacuum distillation, or backflushing with steam, air, or a liquid solvent (e.g., carbon disulfide).

If merely the detection of a specific gas is required, the concentrated adsorbed sample may be treated chemically using chemical spot test methods or specific gas identification tubes.

In some cases, the adsorbed gas may be aqueous colorformed or titrated following extraction of the exposed adsorbant. Filtration is suggested to remove adsorbant particles.

Separation of the Pollutant Gas by Freezeout Techniques

If sample air is drawn through a series of collection coils or chambers, each at a progressively lower temperature, the gas-phase air contaminants are condensed and collected in those chambers whose temperatures are below the boiling point

Table 6 Various Freezing Thermostats for Use in Freezeout Sampling

Liquid or "Blended Puree"	Temperature (°C)
Ice–water	0
Ice–salt (NaCl)	−16
Acetone–dry ice	−80
Liquid air	−147
Liquid oxygen	−183
Liquid nitrogen	−195

of the respective contaminants. Temperatures of these cryoscopic containers are controlled using freezing mixtures such as those shown in Table 6.

The coolant or coolant mixture can be placed in a wide-mouthed Dewar flask and sample air admitted to the freezeout unit through a calcium chloride drying tube to minimize unnecessary water vapor condensations. Such a single stage unit, shown in Figure 19, can be used to collect a gross sample or employed as part of a series to collect a number of pollutant fractions.

Since the efficiency of this condensation method depends on the ability of the collector to contact-cool the pollutant gas, greater efficiencies (70 to 80%) can be approached by packing the collecting chamber with a finely divided high-heat-conductive solid (alumina or metallic beads) and by operating at a flow rate of 0.1 to 0.2 cfm.

After sample collection of 1 to 4 hr, condensed samples must be held at the collection temperature until the time of analysis. Such a sample is quite suitable for gas chromatographic analysis and, if quite dry, can be admitted directly into the chromatograph. Alternatively, the freeze-out trap can be attached upstream of an impinger and a conveniently low airflow used to transfer the concentrated pollutant sample to the impinger, allowing the freezeout trap to gradually return to room temperature or an appropriate subroom temperature.

Collection of Pollutant Gases by Grab Sampling Methods

When either convenience or necessity dictates procurement of a small but representative air sample without recourse to more elaborate air sampling equipment, the use of an evacuated flask or an inflatable bag to capture a static volume of air constitutes "grab" sampling. Such sampling can be either quite nonspecific or specific. In the latter case the evacuated sampler contains a specific reagent solution under partial pressure of air that, on exposure to the sample air and shaking of the flask, yields a direct color-forming or otherwise specific reaction. Quantitation of such a reaction is a simple matter because of the ease of sample volume determination. Here, the exact volume of the grab container can be calculated or measured by volume displacement. Grab samples are usually procured by one of the following techniques.

Evacuated Container

Resealable glass cylinders can be purchased in sizes from 250 ml to 1 liter. These cylinders are attached to a vacuum pump through a manometer connection and evacuated to a desired partial pressure before being sealed with a burner flame as they remain isobaric with the manometer. Such ready-to-use "grab samplers" can be conveniently stored for future use.

An equally ready sampler can be prepared from a spherical flask that is fitted with a glass or plastic cap for rapid and convenient sample handling. Although samplers are fabricated with strong glass walls, it is always advisable to double wrap the outer surface with friction or other webbed tape to ensure against implosion hazard to the air analyst.

Although these sample containers are of known volume, since they are allowed to equilibrate with ambient air at the place of sampling, the volume of air sample must be calculated by correcting the container volume for the atmospheric pressure and temperature at the time of sampling, using ideal gas laws and including the initial partial pressure of the collecting vessel (which should be noted on the cylinder) at the time of evacuation.

Use of evacuated flasks with stopcock fittings is discouraged because of possible sample contamination by the stopcock lubricant.

Displacement of Liquid

A bottle of any convenient size up to 10 to 20 liters is filled with water or a dilute absorbing solution and then used as a collector simply by pouring out and discarding the liquid contained. If the bottle is sufficiently large (>10 liters), the volume of the water or solution remaining on the walls after discard can be neglected in determining the sample volume. Here an error of 0.1% (1

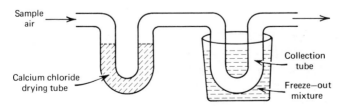

Figure 19. Single unit of a freezeout system.

volume per 1000 volumes) permits no more than 10 cm^3 to remain in a 10-liter flask. If greater analytical accuracy is desired, the liquid is collected and its volume determined after sampling.

Continuous Aspiration

A sample bottle or gas pipette is filled with a grab sample using a syringe or small air pump to draw ambient air through the pipette, eventually displacing its air contents with sample air. Here, it will be necessary to admit at least 10 air changes through the collecting vessel before it can be considered filled with sample air.

Theoretically, no amount of air changes will succeed in removing all of the prior air content. However, a good rule of thumb is 10 to 100 air changes, which might require 3 to 5 min of squeeze bulb operation or 1 to 2 min of pumping. One should remember that the air-moving device should always draw air through the collecting vessel and thus should always be downstream of the collector (i.e., air must enter the collecting vessel before entering the pump or air-moving device).

Inflation of Flexible Containers

The use of plastic sampling bags has for some time been the object of a great deal of study to determine if these very convenient air samplers either adsorb or react with any portion of the sample collected. Mylar bags have been found to be generally satisfactory for the collection of aliphatic and aromatic hydrocarbons, ozone, SO_2, and NO_2. Saran and other materials, including polyester and laminated aluminum films, have been used for the less-reactive gases (e.g., aliphatic hydrocarbons), but are not recommended for any prolonged storage ($> \frac{1}{2}$ hr) of the more-reactive samples. As an added precaution, bags containing pollutant gas samples should be protected from direct sunlight when possible to minimize gas–gas or gas–surface interface reactions.

At the time and place of sampling, such bags are filled by either using an aspirator bulb or, preferably, inflating the bag manually by drawing the bag open with the fingers. It is usually not necessary to know the sample volume, since it is necessary only to procure "enough" sample for future in-laboratory analysis of the contaminated air collected.

SAMPLING PARTICULATE AIR POLLUTANTS

Particulate pollutants are divided generally into dust that settles in air and dust that remains suspended as an aerosol. The physical consideration determining into which class a particle falls is the particle diameter.

Sedimentation, the ability of particles to settle, is defined by a relationship

Table 7 Settling Rates of Various Particles at 18°C

Particle Diameter $(\mu)^a$	Terminal Settling Velocity (ft/hr)
0.01	0.000036
1.0	0.360
5.0	9.24
10.0	36.0

^{a} Assuming particles of specific gravity 1.

that involves the terminal settling velocity of particles falling under the influence of gravity. This relationship, called Stoke's law, allows the analyst either to predict relative settling velocities (V_t) of various particles knowing their diameter (d) or to compute the diameter knowing the time in which a particular aerosol settled from a known elevation.

Assuming that the particles are spherical and that a constant bouyancy is exerted by air, then

$$d = \frac{18_\eta V_t}{(d_p - d_f) g}$$

where η = the viscosity of air (poises)
  d_p = density of the particle (grams per cubic centimeter)
  d_f = density of air (grams per cubic centimeter)
  g = acceleration due to gravity (980 centimeters per square second).

Solving this equation for the settling rates of particles gives the results over a range of diameters shown in Table 7.

From the data in Table 7, it is obvious that the actual settling ability of very small particles approaching molecular diameter is so small that, for practical purposes, such particles remain suspended in air and are even difficult to remove by air washing, as occurs with rainfall.

As a matter of working definition, particles larger than 10-μ diameter are usually thought of as "settleable" while those of a smaller diameter are referred to as "suspended."

Instruments designed to collect either class of particulates are ordinarily chemically passive physical collectors whose function is merely to permit measurement of the collected material without regard to the composition. Generally, the particulates that will be encountered include various mineral

Dust Sampling by Gravity Settling (Dustfall)

Particles generally larger than 10 μ in diameter, which are known to settle from air and collect on horizontal surfaces, can be sampled merely by placing an open container in an outdoor area that is free from overhead obstructions. These collectors are ordinarily constructed of polyethylene, glass, or stainless steel, since the inside walls must be inert to atmospheric oxidative flaking, which would contribute to sample weight. In addition, identical dustfall containers should be employed in the same sampling network or where a comparison of results will be made.

Simple support of the container can be achieved by fixing 3 to 4 dowel rods or rigid wire pins upright in a flat base or table equidistant from the center and following the circumference of the cylindrical container (see Figure 20).

The container is usually slightly more than twice as high as its diameter at the base. This design inhibits the reentrainment of contained dust because of spurious drafts or spatter due to heavy rain.

A typical dustfall container is constructed of plastic and is 8 to 13 in. high and 4 to 6 in. in diameter at the base with a slight inward taper of the walls from top

Figure 20. Simple dustfall collector.

to bottom. Occasionally, water containing algicide (1 mg $CuSO_4$/liter water) is added in the amount of no more than 500 ml to facilitate retention of particles, where known error from loss of particles may occur as a result of gusting winds. Determination of the extent of such errors at a particular sampling site follows from placement of two containers side by side for equal sampling times. Differences of no greater than 10% would be expected to arise from normal settling variances.

In sampling rather large areas, such as entire communities, it is common to employ at least one dustfall container for every 10 mi^2. On the other hand, when dustfall sampling is intended to measure the effect of a given industry or industrial complex, containers may be placed as close as a few hundred feet apart.

Site Selection

Factors determining the selection of a suitable site can be found in specifications published by either the Air Pollution Control office of the EPA[8] or the American Society for Testing Materials.[9] The ASTM standard includes a number of methods for characterization of the dust sample.

General considerations in site selection are:

1. The site should be free from overhead obstructions and away from interference by local sources, such as an incinerator or chimney.

2. The mouth of the dustfall collector should be no less than 8 ft and no more than 50 ft above ground level, with a standard height of 20 ft as a recommended elevation.

3. When sampling in urban areas, the dustfall collector should be set no less than 10 stack lengths from an operating smoke stack and no closer to a vertical wall than the distance that provides a 30° angle from the sampler to the top of the wall or roof.

Procedure for Dustfall Sampling

Apparatus

DUSTFALL COLLECTOR. A light polyethylene jar such as a refrigerator juice container of about 1 to 2 gal capacity and equipped with a press-on lid or similar cap serves as a suitable collector. Metallic containers should be avoided because of possible contamination of the sample by rust or corrosion. A support assembly can be homemade, as in Figure 20, or purchased commercially.

SIEVE. Twenty-mesh brass or stainless steel.

BUCHNER FUNNEL AND FILTRATION FLASK. Recommended 12.5-cm-diameter filter surface.

STEAM OR WATER BATH. Positions to heat 4 to 12 evaporating dishes.

DRYING OVEN. To maintain temperature at 105°C.

WEIGHING BOTTLES. Tall form, 6 cm × 3 cm.

ASPIRATOR. Capable of producing filtration vacuum.

Collection of the Sample

Thoroughly scrub inside walls and bottom of the collecting jar. During summer months, add a known volume (e.g., 500 ml) of distilled water containing 1 ppm (1 mg/liter) of copper sulfate to inhibit growth of algae. Transport the container to dustfall sampling site, remove lid, and mount in support assembly. After exposure of one calendar month, plus or minus a few days, cover the collector with press-on lid and return to the laboratory. Record exact number of days exposed.

Total Dustfall Analysis

Inspect the contents of the sampler and remove such extraneous objects as leaves, twigs, or insects. Perform a preseparation of other large objects by passing the contents through the sieve and return filtrate to the original container. If the container is dry on return to the laboratory, at least ½ liter of water must be added to facilitate the filtration step.

Using a rubber-tipped glass spatula, scrub the inside walls of the collector to loosen dried particulate. Then, with the aid of the spatula and a wash bottle, quantitatively transfer a portion of the contents into a tared 250-ml evaporating dish or beaker. Place the dish on the water or steam bath, replenish as necessary until all liquid is transferred, and continue evaporation of the sample to a moist residue. Complete the evaporation at 105°C in a drying oven; desiccate and weigh to the nearest milligram.

Analysis of Soluble-Insoluble Fractions

Prepare and prefilter the sample as described in the section Total Dustfall Analysis. Do not return the filtrate to the original container. Instead, use a graduated cylinder, adjusting the volume to 1 liter by means of either evaporation or addition of distilled water. Heat the adjusted volume to near boiling to ensure dissolution of any soluble compounds. Cool and quantitatively suction-filter the contents through the Büchner funnel assembly containing a tared Whatman 41H filter paper. Fold the filter loosely, place it in the weighing bottle, and oven-dry at 105°C for 1 hr. Cool the weighing bottle in a desiccator and weigh to the nearest milligram. Report this weight as *insoluble solids.*

Quantitatively transfer the filtrate from the previous separation to a 250-ml evaporating dish. Evaporate as described in the section Total Dustfall Analysis and report the weighed residue as *soluble solids.*

Calculations

Dustfall solids can be expressed in any of the following units:

Grams per square meter per month (30 days).
Pounds per thousand square feet per month (30 days).
Tons per square mile per month (30 days).

The weight of total solids (W_t) should equal the sum of the weights of soluble solids (W_s) and insoluble solids (W_i).

$$\text{total dustfall } D_T = \frac{W_t \times 30 \text{ days}}{A_c \times N_d}$$

where W_t = weight of total solids (grams)
A_c = area of the mouth of the collecting jar (square meters)
N_d = number of days exposed.

To convert grams per square meter per month to tons per square mile per month, multiply by 2.85.

Interpretation of Results

The retention efficiency of the dustfall jar is thought to be improved by the addition of algicide–distilled water in the warmer months or isopropanol (antifreeze) water in winter. However, an evaluation of wet and dry collection by Nader[10] revealed that these approaches are in rather good agreement, allowing for the general reproducibility of 25% for identical side-by-side dustfall collection. Similar results can be found in the works of Sanderson,[11] Stockham,[12] and Lucas.[13]

In areas where unusually heavy rainfall may cause the sampler to overflow, the mouth of the jar can be reduced by fitting with a plastic template containing an opening cut to one-half to two-thirds of the diameter of the original sampler opening.

In addition to weight of soluble and insoluble solids, chemical analysis can be performed on the residue for:

Total sulfates (chemical precipitation).
Chlorides (chemical precipitation).
Combustible material (ignition to constant weight).
Bituminous tars and organics (solvent extraction).
Metals (solution colorimetry or atomic absorption).

The major advantage of dustfall sampling is probably the ease of procurement of 1 to 5 gram of weighable sample, on which a number of chemical and physical (microscopic) analyses can be performed. In addition, the method is simple and inexpensive and requires no electrical power or moving parts.

Additional advantages include the ability to:

Collect dust that is representative of a given industry or community.
Detect process changes of a given industry.
Survey a community to determine areas of high versus low levels of dust pollution.

Disadvantages include lack of precision and inability to distinguish episodes of peak dustfall due to integration of the total sample weight over the entire (up to 30 days) sampling time.

Dust Sampling by Filtration (Suspended Particulate)

As mentioned earlier, the class of settleable particles measured by dustfall collectors represents only a portion of airborne dust. Windblown and groundblown particulate having a particle size of less than 10 μ in diameter, according to Stoke's law, tends to remain entrained in an airstream. Such particulate is referred to as "suspended particulate." Since this class of air contaminant dust is respirable, it is certainly more harmful to man than the larger-diameter settleable particulate. Consequently, it is necessary that air be treated by sampling procedures that discriminate between these two varieties of pollutant dust.

Choice of Filter Media

In the selection of a filter medium, attention must be given to the following:

1. Type of particulate to be collected. Fine particles collected under conditions of extremely high humidity may absorb surface moisture. Moisture can blind a normally hygroscopic filter, causing decrease in flow rate as well as injury to the air pump. Furthermore, since particulates may be either solid or liquid, high-molecular-weight organics may deposit as a film, with resultant similar plugging.

2. Chemical nature of the filter. Some organic particulate may dissolve certain synthetic filters instantaneously, while gases such as NO_2 or SO_2 may attack or decompose some membrane filters. For example, glass-fiber filters should not be used to collect particulate on which analysis for silica is to be performed.

Eventual Recovery of the Sample Particles

If only total sample weight is required, a dense filter mat can be used. However, such a filter impedes the later recovery of sample particles for chemical, microscopic, or particle-size analysis. If only microscopic examination is intended, a filter that collects particles on a flat two-dimensional surface is desirable. Also, the transparency of the filter when immersed in a given solvent is a useful property in microscopic examination of the sample. Here, Pyrex glass fibers disappear in toluene, while cellulose acetate filters are transparent in xylene. Such solvents, of course, would be useless in examination of most highly nonaqueous soluble organic particulates.

Size Range of Sample Particles

Most air sampling filters collect particles to a lower size limit of 0.1 to 1.0 μ diameter. Regardless of type, their efficiency for collection of smaller particles improves as the sampling progresses because of increasing particulate matting within the filter matrix. Placement of several filters in series may increase slightly the amount of sample collected per cubic foot of air. However, the best use of this technique is to follow a coarser filter with a finer filter so that a wider range of particle size is collected.

Actual collection efficiency of filters is a function of a number of variables:

1. Relative humidity of the ambient air.
2. Particle size.
3. Particle shape and surface roughness.
4. Degree of agglomeration.
5. Static charge of the particles and filter.
6. Contact time (thickness of the filter and rate of airflow).
7. Ambient temperature.
8. Chemical properties of the filter material.

Amount of Sample Required for Analysis

Since most (although not all) analytical methods require an initially weighable sample, the lower limit of sample size is often determined by the sensitivity of the analytical balance available. A general rule of thumb is that enough sample air must be collected to provide at least 1 to 10 mg sample on a filter matrix. Thus if the approximate normal air concentration of a particular contaminant is known, determination of the estimated air volume to be sampled, and thus the corresponding sampling time, are relatively simple.

If particulate calcium fluoride is to be sampled (10 mg desired sample weight) and the usual air concentration is known to be in the range of 1 to 50 $\mu g/m^3$, the required volume of air is

$$\frac{10,000 \ \mu g}{1 \ \mu g/m^3} = 10,000 \ m^3 = 351,000 \ ft^3$$

Thus assuming 100% filter collection efficiency, a sampler that filters air at a rate of 60 cfm must operate for a total time of

$$\frac{351,000 \ ft^3}{60 \ cfm} = 5850 \ min \ (97.5 \ hr)$$

Temperature of Particles or Their Surrounding Air

If a sample of rather high-temperature air (e.g., 110°C), such as might be found in the proximity of oxidation or heat-treating processes, is to be taken, a temperature-resistant filter (e.g., glass or polyolefin fiber) should be used in preference to paper or a polyester-type synthetic material.

Classes and Properties of Filters

Some knowledge of the general types of filter material is essential for proper selection of a filter for a given airborne particulate.

Paper Filters

Ordinary laboratory paper filters are suitable for the collection of aerosols, especially when the metal ion content of the particulate is to be determined and low flow rates can be tolerated. The high pressure drop across paper (cellulose) filters restricts their use to the sampling of rather concentrated atmospheres (>250 $\mu g/m^3$) or in the vicinity of a high-level source of particles, such as a steel mill or powder storage area. Also, because of their low metal ion blank, such filters have found most successful application for air sampling within industrial plants rather than ambient air sampling.

Special types of these cellulose filters are shaped to a thickness of about 1 in. through concentric fluting to provide as much as 80 to 100 in.^{2} of filter surface across a 4-in.-diameter airstream.

When a rather refractory particulate (e.g., silica or limestone) is to be collected, cellulose filters can be conveniently ashed to yield a dry weighable sample, which is ready immediately for further chemical analysis by atomic absorption or colorimetry.

Membrane Filters

Compaction of a dry gel of a cellulose ester into a mat composed of tiny 1 to 2 μ conical pores yields a filter matrix that allows moderate flow rates of sample air through the pores while effecting particle separation by entrapment of particles on the inner pore walls as the sample air spirals through the pore itself. Particles are thus quite viewable microscopically as well as accessible for chemical analysis by aqueous extraction. In addition, the filters can be rendered transparent by the addition of an immersion oil having a refractive index of 1.5, so that the retained particulate becomes quite visible from any viewing angle.

Efficiency of collection is $>99\%$ for submicron aerosols to a lower limit of 0.2 μ.

Unfortunately, the acetate matrix often solubilizes organic particulate, with subsequent collapse of the conical walls of the three-dimensional pore structure.

Although permissible flow rates are higher than for paper filters, flow rate is limited to 10 to 20 cfm.

Inorganic Fiber Filters

When high flow rates (e.g., 20 to 60 cfm) are required to collect a large amount of sample at little sacrifice of collection efficiency, glass-fiber filters provide a means of air filtering up to 1 g of particulate in a relatively short time (24 hr). These filters contain no organic binder and are probably the most suitable matrix for use in collecting organic aerosol, which is then easily solvent-extracted.

Metal ion content of the glass fibers requires careful determination of a blank when this filter is used to collect particulate for subsequent metal ion determination. Here, 5 to 10 clean papers should be acid-extracted or, if laboratory space and equipment is limited, a representative blank can be determined by extracting multiple outer edge and center portions from several filters to yield the equivalent of 1 to 2 filters. Also, some difficulty may be encountered in recovery of 100% of the particles, some of which resist extraction and remain enmeshed in the dense filter matrix.

Sampling Suspended Particulate by High-Volume Filtration (The High Volume Sampler)

Principle of Operation

Stoke's law offers the basic approach to suspended particulate collection.

When operated in virtually quiescent air, the geometry of the high-volume sampler (see Figure 21) employs the sloping roof of the shelter as a means for causing air entering the sampler under the eaves of the roof to change direction at least 90° before entering the horizontal filter.

Particles that remain entrained in the air sample prior to horizontal filtration have, in so doing, satisfied the definition of truly suspended dust—dust that is not subject to settling under the influence of gravitational force.

Air Sampler Operation

The sampler, commonly known as a "Hi-Vol," is a vacuum cleaner-type motor that is used to draw sample air through a filter area. The filter can be:

1. A 4-in.-diameter circular glass-fiber disk.
2. A 4 × 4 in.^{2} glass-fiber mat.
3. A standard 8 × 10 in. glass-fiber mat.

The filter most frequently employed is the 8 × 10 in. mat, which allows collection of an air sample at a rate from 40 to 60 cfm over a nominal sampling period of 4 to 6 hr and a normal sampling period of 24 hr. These conditions permit the sampling of from 65,000 to 75,000 ft^3 of ambient air, with consequent extraction of about ½ g of suspended particulate (aerosol). This provides quite a substantial weight of sample, which greatly simplifies subsequent chemical or physical analysis.

Figure 21. High volume (hi-vol) air sampler (Courtesy of Research Appliance Co., Allison Pk., Pa.; manufactured by General Metal Works, Cleves, Ohio)

The motor is usually started and stopped by a simple clock timer, and the duration of sampling is measured by an elapsed time meter that is placed in series with the Hi-Vol motor. Starting and finishing times are at the discretion of the operator, although the EPA recommends starting and finishing from midnight to midnight—24 hr over the same calendar day. The National Air Sampling Network operates such samplers over the entire country on scheduled 24-hr intervals every sixth day. This schedule permits a sampling of every day of a week over a period of 7 weeks and is a recommended approach for long-term community air sampling. On the other hand, short-term studies to determine day-to-day variation in particulate levels may require continuous daily 24-hr sampling. Fulfillment of the aims of such a program, without manual changing of filters at midnight, is possible by:

1. Using one sampler and manually changing the filter at noon or some other convenient hour rather than midnight.

2. Using two Hi-Vol samplers side by side, with a timer switch set to start a second motor as the first stops at midnight. The exposed filter can then be changed conveniently during the daylight hours while the second 24-hr sample is being collected.

An airflow measuring device, such as a rotameter, is used to record the volume of sample air passing through the filter for later use in calculating the mass of particulate per static volume of air (micrograms dust per cubic meter of sample air). The rotameter is usually first calibrated to the individual Hi-Vol sampler (see calibration) and then used to measure the "initial airflow" through the clean filter mat. At completion of a 24-hr (or other) sampling time, another rotameter reading is made on the exposed filter, and the average flow is determined.

Procedure

The succession of operations involved in collecting a sample of aerosol particulate using this device (see Figure 22) can be summarized as follows:

1. Install a single sheet of tared, desiccated, glass-fiber filter on the screen support of the Hi-Vol sampler no more than 24 hr before intended start of sampling. (Earlier placement of the clean paper can result in as much as 5 to 10 mg deposition before start of sampling from air currents through the sampler).

2. Using a rotameter that has been calibrated to measure the actual airflow through the sampler, record the initial sample airflow with the sampler "on" and the fresh filter in place.

3. Set the on–off time to start at midnight (or some other preselected hour) on the intended sampling calendar day.

4. Return within 1 day after sample collection and carefully remove the exposed filter without tearing or touching the collecting surface. High winds or inclement weather occasionally make it necessary for the operator to shield the sampler with his body or clothing while transferring the filter.

5. Fold the filter twice along either its width or length and place in an envelope that may be used as a labeling surface on which to record pertinent information. An example of suggested labeling for a 5 × 7 envelope is shown in Figure 23.

Filters Used in High-Volume Air Sampling

The use of common laboratory analytical-grade paper filters is not recommended for the sampling of ambient air. The large-diameter cellulose fibers,

Figure 22. Close up of sampling head, high volume air sampler (Courtesy of Research Appliance Co., manufactured by General Metal Works, Cleves, Ohio)

which are irregular and can have flattened walls, do not provide good aerodynamics for the collection and recovery of very small ($< 0.5\ \mu$) particles. In addition, such paper filters often do not possess the mechanical strength to withstand airflows of 40 to 60 cfm, which result in a pressure differential across the filter of 10 to 20 in. of mercury.

A filter mat prepared from interwoven glass fibers provides a dense porous medium through which an air stream must change direction in a random fashion, allowing the entrained particles to collide with and become entrapped on filter fibers.

The standard 8 × 10 in. glass-fiber filter is designed to collect up to 1.0 to 1.5 g of particulate. Operation beyond the 24-hr recommended sampling time may result in collecting too large a sample, which may "blind" the filter, cause damage to the sampler motor, and result in possible loss of sample due to wind blowing of loose excess particulate during filter retrieval.

Ordinarily, collection efficiency of the glass-fiber filter is thought of as >99.9% for particles in the size region of 0.1 to 10 μ diameter.

Among the few disadvantages encountered in the use of the glass-fiber filter are difficulties in recovering particulate suitable for determination of size distribution (due to fracture of large particles on collision with fibers) and interference in determination of airborne metals (due to the metal ion content of the filter). Both of these shortcomings can be avoided if one employs a two-filter system in which a cellulose acetate (membrane) filter is placed on the surface of the glass-fiber filter, which is then used only as a means of support. Particulate collected on the membrane filter can be recovered by dissolving the plastic membrane in ethyl acetate or other solvent. The particle-size distribution can be determined either by liquid counting in the solvent or by evaporating the solvent to allow microscopic particle-size determination. Since a greater pressure dif-

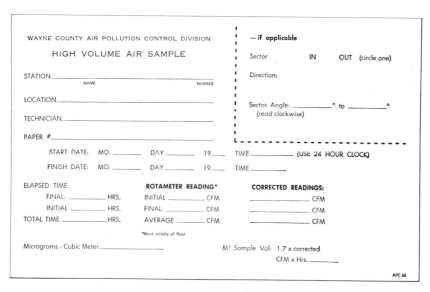

Figure 23. Example of a printed label for use in identifying stored envelopes containing air sample filters.

240   Principles of Air Sampling

ferential exists across such a double-filter arrangement, lower sample airflows of 5 to 20 cfm are in order.

Before a filter can be used to collect a particulate sample, it must first be dried and weighed. Although oven-drying at 110°C permits rapid filter preparation, subsequent oven-redrying following sample collection tends to volatilize most organic or hydrocarbon particulate collected. Consequently, the EPA has recommended 24-hr equilibration at atmospheric relative humidity as filter conditioning prior to weighing of the clean filter and reweighing of the exposed filter. Although a laboratory atmosphere is suggested, almost any reproducible atmosphere of low humidity, such as a laboratory dry box or desiccator, can be used to equilibrate the filter.

Calculations

The most common expression of high-volume air sampling data is in micrograms of suspended particulate per cubic meter of air, representing a 24-hr sampling period. Using typical results of such a sampling, an example calculation would proceed as follows:

VOLUME OF AIR.   Sample airflow, corrected rotameter reading:
   Clean filter   56 cfm
   Filter after exposure   52 cfm
   Average   54 cfm

Therefore,
$$\frac{54 \text{ ft}^3}{\text{min}} \times \frac{1440 \text{ min}}{24 \text{ hr}} = \frac{76{,}760 \text{ ft}^3}{24 \text{ hr}}$$

and
$$\frac{76{,}760 \text{ ft}^3}{24 \text{ hr}} \times \frac{0.0283 \text{ m}^3}{\text{ft}^3} = \frac{2172 \text{ m}^3}{24 \text{ hr}}$$

WEIGHT OF PARTICULATE.   Tare weight of filter:
   Before exposure   3.417
   After exposure   3.925
   ─────────────────────────
   0.508 g or 508×10^3 µg

SUSPENDED-PARTICULATE CONCENTRATION

$$\frac{508 \times 10^3 \text{ µg}}{2172 \text{ m}^3} = \frac{234 \text{ µg}}{\text{m}^3}, \quad \text{per 24-hr sample}$$

Calibration of the Hi-Vol Flow System

PRINCIPLE.   The rate of sample airflow is simply a function of the speed of the motor, which is, in turn, dependent on electrical power. Thus as shown in

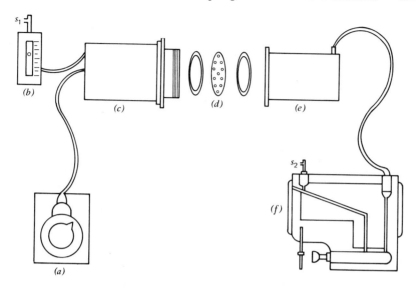

Figure 24. Exploded view of critical calibration using inclined manometer and orifice plates. (a) Rheostat; (b) Rotameter; (c) Hi-Vol motor; (d) orifice plate; (e) orifice meter; (f) inclined manometer. (Connect s_1 to s_2 using flexible tubing)

Figure 24, it is a simple matter to vary the sample airflow by using a rheostat (a) to set the speed of the motor (c) while measuring the pressure drop across the orifice meter (e) by means of an inclined manometer (f) and an orifice plate (d). A rotameter (b) reading is made at airflows that cause manometer readings of 1.30 through 4.90 in. of water pressure differential. The inches of water are converted to cubic feet per minute by means of a graph supplied with the standard orifice plate. A plot is then prepared of rotameter readings recorded at each measured increase in airflow cubic feet per minute (Table 8) and a line drawn to connect points as shown in Figure 25.

The rotameter thus calibrated should be marked with the same identification number as the corresponding Hi-Vol sampler and a record kept of the number and the calibration chart.

Repeated referral to such a chart can be minimized if a rheostat is used at each Hi-Vol sampler to set the flow, as measured by rotameter, to some predesignated rotameter value (e.g., 50 or 60), which would then correspond to a known cubic feet per minute (e.g., 54 or 63). As a matter of convenience, an entire sampling network can be adjusted to the same initial flow rate, provided that motors are of comparable operating efficiency.

Recently, the Environmental Protection Agency has published a standardized procedure[14] to be used in calibrating high-volume air samplers under the

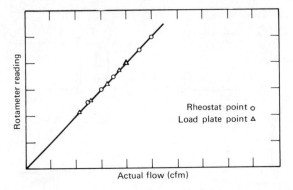

Figure 25. Calibration curve for sample air flow through Hi-Vol system (see Table VIII).

reference method for Determination of Suspended Particulates in the Atmosphere.

Here, an orifice meter is situated ahead of the Hi-Vol motor to monitor the airflow, and a series of perforated load plates are placed, one at a time, between this calibration unit and the Hi-Vol motor, as shown in Figure 25.

CALIBRATION PROCEDURE. A working procedure for this calibration is as follows:

1. Assemble the calibration apparatus as shown in Figure 25.

Table 8 Example of Data for Calibration of Rotameter to Hi-Vol Samples

Rheostat Setting	Loadplate Number	Manometer Reading (in. H_2O)	Actual cfm	Rotameter Reading
93.	—	8.35	50.	50.
83.5	—	6.50	45.	45.
75.	18	4.90	40.	40.
69.	13	4.10	37.	37.
65.	—	3.58	35.	35.
59.	10	3.00	32.	32.
55.5	—	2.62	30.	30.
47.	7	1.90	25.5	25.5
46.5	—	1.80	25.	25.
40.	5	1.30	21.5	21.0

2. Turn on the variac (rheostat) (*a*) and slowly increase the voltage until the desired differential pressure (e.g., 4.90 in. of water for 40 cfm using the 18-hole perforated loadplate) is achieved. Use the inclined manometer (*f*) to measure the static pressure.

3. When a constant pressure (e.g., 4.0 to 5.0 in. of water) is achieved, set the rotameter to read exactly 40.0 or 50.0 cfm and dope the adjustment nuts on the rotameter so that any disturbance will be easily recognized.

4. Replace loadplates one at a time (i.e., 13, 10, etc.) and reset to successively lower differential manometer readings (i.e., 4.10, 3.00) by again slowly decreasing the variac voltage. Record the corresponding rotameter reading as shown in Figure 26.

Based on a collaborative study[15] of the high-volume method, the use of resistance or load plates for calibration is recommended over the voltage variation (rheostat) method. Here, the two approaches have been found to be equivalent at high flow rates (i.e., approaching 60 cfm).

Maintenance

A sampler that operates 24 hr one or more times per week requires a 6-month servicing check in which motor brushes should be replaced and the electrical system checked for loose or worn wires. At least once per year, a calibration check should be performed; the rubber gasket on the filter frame should be changed at this time. In the event of an unscheduled repair or maintenance of the Hi-Vol, a recalibration of the instrument is necessary to ensure measurement accuracy.

The Soil-Haze Particle Sampler (Tape Sampler)

Although the dustfall and high-volume air samplers are limited to sampling periods of 24 or more hours because of the need to collect an analytically weighable sample, the tape sampler permits aerosol particulate sampling that can be varied from 5 min to 3 hr. Rather than an actual particulate air loading, the tape sampler measures the degree of air soiling or the density of the haze present in a package of air.

Principle of Operation

Ambient air containing suspended particles is drawn into the sampler and through a limiting nozzle across which a length of filter tape is held in a track. Particles are deposited on the tape, and the tape is advanced at predetermined intervals (e.g., ½ hr, 2 hr) to an area of clean tape where the next sample of aerosol is collected.

Spots so obtained represent soiling during the sampling period. They are usually evaluated by measuring the amount of light that can be transmitted

High volume air sampler

Orifice calibration summary sheet

Hi–Vol # _____   Date _____

Rotameter # _____   Name _____

Hi–Vol location _____

Differential manometer

Load plate No.	Inches H$_2$O	cfm	Rotameter cfm
18	4.90	40*	
13	4.10	37*	
10	3.00	32*	
7	1.90	25.5	
5	1.30	21.5	

*If differential manometer cfm does not equal rotameter cfm, then a calibration curve must be prepared.

Maintenance: _____

_____

_____

_____

_____

_____

_____

Figure 26. Example of data chart for calibration and maintenance of high volume air sampler.

through the spot. This measurement can be accomplished by "reading" the transmittance of each spot in succession and determining the optical density.

$$\text{optical density (OD)} = \log \frac{1}{\text{transmittance}} = \log \frac{I_0}{I}$$

where I_0 = the transmittance of clean tape, 100%.
I = measured transmittance of soiled tape, usually less than 100%.

Calculation

Data are usually reported in COH's (coefficient of optical haze per 1000 linear feet) of sample air, where COH is a soiling haze index and is defined as the quantity of light scattering solids producing an optical density of 0.01 when measured by light transmittance.

The package of air sampled can be envisioned as a cylinder of the same cross-sectional area as the tape spot and a "length" dependent on the sampling time. Thus COH is calculated by using the optical density, the area of the spot, and the sample volume.

$$\frac{\text{COH}}{1000 \text{ ft}} = \log \frac{I_0}{I} \times \frac{A \times 10^5}{V}$$

where A is the area of the spot in square feet and V is the sample volume in cubic feet or the product of the average sample flow (cubic feet per minute) and the sampling time (minutes).

For example, what is the COH per 1000 ft if a sample spot of 1.0-in. diameter is found to have a transmittance of 50% after 2 hr of sampling at an average flow rate of 0.2 cfm?

Here,

$$A = \frac{\pi d^2}{4} = \frac{\pi \times l^2}{4} \times \frac{l \text{ ft}^2}{144 \text{ in.}^2} = 5.45 \times 10^{-3} \text{ ft}^2$$

$$v = \frac{0.2 \text{ ft}^3}{\text{min}} \times 120 \text{ min} = 24 \text{ ft}^3$$

$$\log \frac{100}{50} = \log 2 = 0.301$$

Substituting,

$$\frac{\text{COH}}{1000 \text{ ft}} = 0.301 \times \frac{5.45 \times 10^{-3} \times 10^5}{24} = 8.3$$

Interpretation of Soil-Haze Data

The soil haze index (COH) represents, of course, an optical property much different from the gravimetric expressions of micrograms per cubic meter or grams per square meter that have been discussed previously for particulate measurements. Consequently some interpretation must be made for various levels of soiling haze index to produce a useful air measurement.

A typical rating system based on visibility is shown in Table 9.

In addition, a number of attempts have been made to relate COH indices to levels of suspended particulate as measured by high-volume air sampling.[16-18] The relationship

$$\frac{\mu g}{m^3} = 240 \, (COH)^{0.91}$$

has been developed by the Ontario Department of Energy and Resources Management, Air Management Branch, where the correlation coefficient for the above relationship is 0.70.

Another expression for the COH, which employs a more convenient scale of units has been developed by the Wayne County Air Pollution Control Agency, Detroit, Michigan. Here the acronym MURC (Measure of Undesirable Respirable Contaminants) is used to express soiling haze in values that more closely approximate temperature and relative humidity scales. The MURC equation is

$$MURC = 70(COH)^{0.7}$$

Using this equation, a COH of 0.5 yields a MURC of 43.

Although correlation between MURC and suspended particulate has not proved consistent, a general trend of correspondence is evidenced by Figure 27.

In respect to such attempts to compare particle concentration in weight units to the degree of decrease in visibility, such variables as density and size distribution of the particulate, together with their coloration and translucence, render discernment of a true standard correlation difficult, if not impossible. Probably the best such relationship can be achieved between side-by-side measurements of the same type of suspended dust over a long enough time period to permit evaluation of the effect of seasonal variation.

Table 9 Evaluation of Soil-Haze Indices

COH	Soil-Haze Pollution
0.0–1.0	Light
1.1–2.0	Medium
2.1–3.0	Heavy
3.0	Very heavy

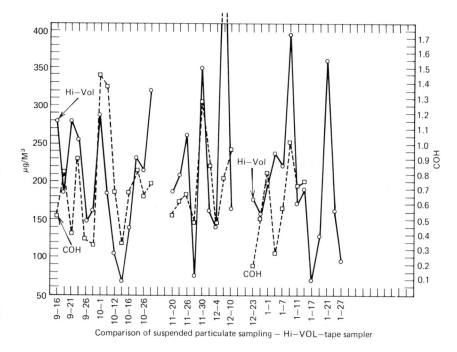

Figure 27. Comparison of suspended particulate sampling by High-volume and tape sampler at a single sampling site.

Collection of Nonparticulate Pollutants Using Reactive Tape

Extension of the use of the tape sampler to the collection of gaseous pollutants is possible through the use of chemically reactive tape and an inert filter plug to prevent the entrance of particulate. Tapes impregnated with lead or mercury salts darken upon exposure to sample air that contains H_2S, while organic dyes and other reactive media can be substituted for standard Whatman No. 4 cellulose paper to collect fluoride and other halides.

Calibration and Maintenance

The two basic variables in soil-haze sampling are airflow and specific transmittance. Transmittance is calibrated by use of film strips of specific known opacity. Airflow, which usually requires more frequent attention, is calibrated using a rotameter that has been standarized against a standard dry gas meter.

A typical automatic instrument operates on a round-the-clock basis. Sampling intervals of 2 hr at a sample flow rate of 0.25 cfm result in a sample volume of 30 ft^3.

At such operating conditions, it is necessary to maintain $<7\%$ deviation in airflow to report the COH to the nearest tenth. Consequently, a weekly flow check is recommended.

Maintenance consists of periodic replacement, as needed, of the light source and cleaning of the internal rotameter. Operation of an external rather than internal vacuum pump simplifies frequent (quarterly) replacement of the diaphragm and minimizes instrument wear due to vibration.

Other Approaches to Particulate Sampling

Impingement

Separation of particulates from an airstream by abrupt collision against a flat surface forms the basis for the impinger, as treated previously in the section on Sampling Gaseous Pollutants.

If the surface is submerged in a liquid, the sampler is known as a "wet impinger," and the viscosity of the liquid acts as a deterrent to possible reentrainment of particulate. Dry impingers rely only on impact for particle collection. Although not as efficient for the smaller diameter ($<2\ \mu$) particles, dry impingers permit collection of a sample directly, without filtration or evaporation of a liquid medium. This is important in cases where sample collection is to be followed by microscopic analysis.

Certain dry impingers are designed to collect particulates at low flow rates to prevent particle fracture, while separating the particulates into size groups on the basis of inertia. Such samplers, called cascade impactors, often employ weighable surfaces or, in some cases, microscope slides as the actual collecting surfaces.

Thermal Precipitation

The migration of particles from a zone of heated air to a layer of cold air because of rapid motion of the air around the particle provides the basis for the collection of particulate by thermal precipitation. This behavior is so subtle as to be unnoticeable in air currents of modest temperatures; however, the effect of thermal repulsion is substantial in cases of temperature gradients of as much as 3000 to 4000°C/cm. Indeed, efficiencies approaching 100% are claimed for particles of low specific thermal conductivity (e.g., carbon).[19]

This type of collector is especially valuable when it is necessary to collect particulates without deglomeration or transformation of the in situ particles. Disadvantages include necessarily low sampling rates of 0.02 to 0.2 liters/min.

Electrostatic Precipitation

Application of a potential difference across a sample airflow, so that actual ionization of the gas volume takes place, leads to collisions between particles and ions to produce electrostatically charged particles. Subsequent passage of these charged particles through an electrical field results in their collection at an electrode whose charge is opposite to that of the particle.

Such samplers are usually quite portable and operate by passing a flow of sample air past a negative central ionizing electrode. The resulting negatively charged particulate is collected on the ground or positive sheaf, which also acts as the container for the ionizing electrode. Potential difference between the ionizing and collecting electrode is as high as 25,000 to 30,000 V. Although a sample collected electrostaticlly is agglomerated, it is quite suitable for chemical or microscopic analysis. Ordinarily, these samplers are employed for source sampling or monitoring atmospheres with relatively high particulate loading.

Inertial Centrifugal Collection

When a rapidly moving stream of air is caused to rotate in order to follow a helical path, particles in the air stream tend to follow the most direct inertial path. Therefore, they migrate toward the outside walls of the vortex, from which they can be collected at the end of the sampling period.

Advantages include procurement of a dry, chemically pure sample. The single most important disadvantage is the limit of this method to the particle sizes greater than 5 μ. Ultrarapid centrifugation has been employed with some success to separate the submicron particles.

AIR SAMPLING AND CHEMICAL DETERMINATION OF COMMON INDUSTRIAL POLLUTANTS

Outlines of recommended air sampling approaches for collection and analysis of both particulate and gases are provided in Tables 10 and 11. Sampling rates are calculated from average air concentrations. The concentrations of the most common contaminants are given for several United States cities in Table 12 and 13 to provide easy reference and comparison.

APPLICATION OF STATISTICAL METHODS TO AIR SAMPLING AND DATA EVALUATION

The application of statistics to air analysis generally serves one of two purposes—community air modeling or evaluation of data.

Community Air Modeling

The development of statistical prediction methods for dealing with extremely complex physical environmental systems has always depended on the prior

Table 10 Air Analysis of Common Industrial Particulates

Pollutant	Sampling Apparatus	Sampling Rate (cfm)	Minimum Sample Weight (g)
Silica	Wet impinger, water	1	1–3
	High-volume sampler	50	1–3
Carbon soot	High-volume sampler	50	1–3
	Thermal precipitator	0.01	0.5
Sulfur trioxide	Wet impinger, aq. alkali	1	0.10 mg
(sulfuric acid mist)	Wet impinger, aq. alkali	1	0.01 mg
Zinc oxide	High-volume sampler	50	0.001
(smelter, die casting)	Electrostatic precipitator	2–3	
Flyash	High-volume sampler	50	0.1
(power plant emission)	Centrifugal collector		0.1
Limestone	High-volume sampler	50	0.10
(kilning)	Dustfall	—	0.01
Iron oxide	High-volume sampling	50	0.001
	Dustfall	—	0.001

Analytical Method	Chapter Reference	Other References
X-ray diffractometry Total weight	1	Talvetie and Brewer, *AIHAJ*, **23**; 214 (1962).
Combustion, CO_2		Warner, Saad, and Jackson, *JAPCA*, **22**; 887, 1972.
Microscopic	1, 2	Crozier and Seeley, Tech, Report, N.M. School of Mines, **1**; 20 (1949).
Acidification to remove CO_2, back titration with standard base or sodium tetraborate Titration using barium ion and thorin indicator	1, 3	Commins, *Analyst*, **88**; 364 (1963). Mader, Hamming, and Bellin, *Anal. Chem.* **22**; 1181 (1950). U.S. PHS Method No. 999-AP-11-I1.
Acid extraction Atomic absorption	1	Allen, *Analyst*, **86**; 530 (1961), or Thompson, Morgan, and Purdue, *Analysis Instrumentation*, Instrument Soc. of Amer., Pittsburgh, Pa., 15219.
Total weight Microscopic	1	McCrone, *The Particle Atlas*, Ann Arbor Publ., Ann Arbor, Mich., 1967. Chamot and Mason, *Handbook of Chemical Microscopy*, John Wiley & Sons New York 1963.
Atomic absorption (Ca) X-ray diffractometry	1	David, *Analyst*, **84**; 495 (1960). Warner, Saad, and Jackson, *JAPCA*, **22**; 887, 1972.
Titration Atomic absorption	1	*Methods for Air Sampling and Analysis*, Amer. Public Health Assoc., Wash., D.C. 1972, Intersociety Method No. 12126-02-70T. Fortune and Mellan, *Ind. Eng. Chem. Anal., Ed.* **10**; 60 (1938).

Table 10 (cont'd)

Pollutant	Sampling Apparatus	Sampling Rate (cfm)	Minimum Sample Weight (g)
Chromic acid (vapor metal plating)	Wet impinger, alkali	0.5–1.0	0.01
Cement dust (kilning)	Dustfall High-volume sampler	— 50	0.10 0.01
Gypsum (kilning)	High-volume sampler Wet or dry impinger Dustfall	50 1 —	0.01 0.10 0.10
Road salt	High-volume sampler Electrostatic precipitator Dustfall	50 2–3 —	0.01 0.01 0.10
Lead (smelting, casting, reclaiming)	High-volume sampler	50	0.01
Mercury	High-volume sampler Freezeout trap Electrostatic precipitator	50 1 2–3	0.01 0.01 0.01
Cadmium	High-volume sampler	50	0.01

Analytical Method	Chapter Reference	Other References
Colorimetry	1	Willard and Diehl, *Advanced Quantitative Analysis*, VanNostrand, New York, 1960, p. 230.
Microscopic analysis X-ray diffractometry Total weight	1	McCrone, Drafty, and Delly, *The Particle Atlas*, Ann Arbor Publ., Ann Arbor, Mich., 1969. Warner, Saad, and Jackson, *JAPCA*, **22**; 887, 1972.
Total weight X-ray diffractometry X-ray diffractometry	1	Warner, Saad, and Jackson, *JAPCA*, **22**; 887, 1972. McCone, Drafty, and Delly, *The Particle Atlas*, Ann Arbor Publ., Ann Arbor, Mich., 1969.
Extraction emission spect. (Na), or titration with mercuric chloride X-ray diffractometry, titration X-ray diffractometry, titration	1	Beckman Technical Bulletin, No. 6072C. *Methods for Air Sampling and Analysis*, Amer. Puclic Health Assoc., Wash., D.C. Intersociety Method No. 12203-01-68T, 1972. American Socity for Testing Materials Method D512, Part 23, 22–24, 1964.
Extraction atomic absorption, colorimetry	1	*Methods for Air Sampling and Analysis*, Intersociety Method No. 12128-01-71T, 1972.
Thermal extraction, atomic absorption Atomic absorption Atomic absorption	1	*Analytical Methods for Atomic Absorption Spectrophotometry*, Perkin Elmer Instrument Manual, Method GC-2, March 1971.
Atomic absorption	1	Thompson, Morgan, and Purdue, *Analysis Instrumentation*, **7**; Instrument Soc. of America, Pittsburgh, Pa., 15219.

Table 10 (con'd)

Pollutant	Sampling Apparatus	Sampling Rate (cfm)	Minimum Sample Weight (g)
Beryllium	High-volume sampler	50	0.1
Fluorides	Wet impinger, aq.	1	0.01
	Fritted bubbler, alkali	1	0.01
	Electrostatic precipitator	3	0.01
Poly cyclic compounds	High-volume sampler	50	0.01
Sulfates	High-volume sampler	50	0.001
	Wet impinger	0.5–1	0.001
Nitrate	High-volume sampler	50	0.005
	Wet impinger	0.5–1	0.005
Ammonium	High-volume sampler	50	0.001
	Wet impinger	0.5–1	0.001
	Freezout trap	0.1	—

development of improved methods for weather prediction. The first such attempt to relate pollution level data with meteorological data was probably made by Hilst and Bryan,[21] who, working with 2 to 3 years of collective data from the 52 stations in the National Air Sampling Network (NASN), published a preliminary correlation of meteorological and air-quality parameters for a number of population centers throughout the United States in 1962. For the first

Analytical Method	Chapter Reference	Other References
Atomic absorption	1	Thompson, Morgan, and Purdue, *Analysis Instrumentation*, **7**; Instrument Soc. of America, Pittsburgh, Pa., 15219.
Colorimetric Colorimetric Colorimetric	3	*Methods for Air Sampling and Analysis*, Intersociety Method No. 12202-01-68T, 1972. Megregian, *Anal. Chem.*, **26**; 1161 (1954).
Spectrofluorometric, spectrophotometric	2	Sawicki, et al., *Atmos. Environ.*, **1**; 131–145 (1967). *Methods of Air Sampling and Analysis*, Intersociety Method No. 11104-01-69T and No. 1104-02-69T, 1972.
Colorimetry, methyl Thymol blue	5	Register of Air Pollution Analyses, US PHS Publ. No. 610, Vol. 1, 1958; Vol. II, 1961. US PHS Publ. No. 999 AP-111-I-1.
Colorimetry, 2–4 xylenol	5	Intersociety Committee No. 12306-01-70T.
Colorimetry, sodium phenolate	5	Tetlow and Wilson, *Analyst*, **89**; 453–465 (1969).

time, they demonstrated that prediction of elevated pollution levels from meteorological variables alone is possible.

The chief advantage of this technique is the ability to arrive at a model of pollution increase or decrease for a given community when given a number of meteorological variables. Specifically, conditions of low wind velocity coupled with low inversion level (air ceiling, below which smoke tends to gather) are cir-

Table 11 Air Analysis of Common Industrial Gaseous Pollutants

Pollutant	Sampling Apparatus	Sampling Rate (cfm)	Minimum Sample Volume (ft^3)
SO_2 (fossil fuel burning)	Wet impinger, or fritted aq., $HgCl_2$ + NaCl	0.1–1	1–10
NO_2 (fossil fuel burning)	Wet impinger Grab sample	0.1 —	1.00 0.01
Hydrocarbons (oil refining, fuel burning)	Flame ionization detection Freeze-out trap Grab sample One liter collection bottle or gas pipette	<1 0.1 Batch sample	0.005 10.000 0.200
CO (fuel burning)	Infrared absorption Hopcalite conversion Nondispersive infrared	Batch sample <1 0.2	0.5–10 0.05–1 Continuous
Ozone (sunlight, hydrocarbons)	Chemiluminescent Wet impinger, aq., KI	0.5–1.5 0.1	Continuous 1.5
H_2S	Wet impinger, aq., $CdCl_2$	0.05	0.5

Analytical Method	Sensitivity (ppm)	Chapter Reference	Other References
Colorimetric, West-Gaeke	0.005	3	West and Gaeke, *Anal. Chem.*, **28**; 1816 (1956). *Methods of Air Sampling and Analysis*, Intersociety Method, No. 420401-01-69T (1972).
Colorimetric, Saltzman	0.01	3	*Selected Methods for Measurement of Air Pollutants*, PHS Publ. No. 999-AP-11, C-1, 1965.
Colorimetric, Saltzman	0.01		Saltzman, *Anal. Chem.* **26**; 1949–1955 (1954).
Automated flame ionization	1–5		*Methods of Air Sampling and Analysis*, Intersociety Method, No. 43101-02-71T (1972).
Gas Chromatography	<1	1, 2, 4	
Gas Chromatography	<1		*Methods of Air Sampling and Analysis*, Intersociety Method, No. 43101-01-69T (1972).
Infrared spectrophotometer	1–5		*Methods of Air Sampling and Analysis*, Intersociety Committee Method, No. 42101-03-69T.
Thermometric, gravimetric	1–10	3, 4	Lysyj, Zarembo, Hanley, *Anal. Chem.*, 31; 902–904 (1959).
Automated NDIR	1–5		Beckman Instructions Manual, 1307-C, Beckman Instruments Inc., Fullerton, Calif., May, 1966.
Automated chemiluminescent, ethylene	0.005	2, 3, 4	*Federal Register*, April 30, 1971. *Selected Methods for Measurement of Air Pollutants*, PHS Publ. 999-AP-11, D-1 (1965).
Colorimetric, Saltzman	0.020		
Colorimetric	0.001	3	Jacobs, Braverman and Hochheiser, *Anal. Chem.*, **29**; 1349 (1957).

Table 11 (cont'd)

Pollutant	Sampling Apparatus	Sampling Rate (cfm)	Minimum Sample Volume (ft^2)
Ketones	Fritted bubbler, aq	0.05	1
Aliphatic aldehydes (fuel burning)	Wet impinger, MBTH	0.1	1–2
Formaldehyde (incineration)	Wet impinger, aq	0.1	1–2
Methanol (chemical manufacture)	Wet impinger, aq., or freeze-out trap	0.1	1
Odors (industrial)	Grab sample 250 ml collection bottle, or gas pipette	Batch sample	0.01
CO_2	Wet impinger with sodium butanolate or fritted bubbler with Ba ion	0.1–1.0	1–10
	Ascarite absorption	0.1	1
Acrolein	Wet impinger, ethanolic 4-hexylrecorcinol	0.05–0.10	1–2
Halogenated compounds (chlorine containing pesticides)	100-ml gas collection bottle or gas pipette	Batch sample	0.001

Analytical Method	Sensitivity (ppm)	Chapter Reference	Other References
Colorimetric	1	2	Coulson, *Anal. Chim Acta*, **19**; 284–288 (1958).
Colorimetric, MBTH	0.01	2	Levaggi and Feldstein, *JAPCA*, **20**; 312 (1970).
Colorimetric, chromatropic acid	0.01	2	Smith et al., *Health Lab. Sci.*, **7**; 87 (1970). Sleva and Stanley, *Publ. Health Ser. Publ.*, No. 999-AP-11, H-1, 1965.
Colorimetric, chromatropic acid	0.01	2	*Methods of Air Sampling and Analysis*, Intersociety Method No. 43502-02-70T (1972).
Odor panel	1.0 (ou)	2	American Society for Testing Materials, ASTM Method 01319-67, Part. 23, p. 301, 1972.
Titrimetric, gravimetric	0.1	2	Stern, *Air Pollution*, Vol. II, Academic Press, New York, 1968, p. 108.
Gravimetric	0.1		Furman, *Standard Methods of Chemical Analysis*, VanNostrand, New York, 1966, p. 300.
Colorimetric	0.01	2	Cohen and Saltzman, *Pub. Health Ser. Publ.* No. 999-AP-11, G-1, 1965. *Methods of Air Sampling and Analysis*, Intersociety Method No. 43505-01-70T (1972).
Gas chromatography electron capture detection	0.001	2	Saltzman, Coleman, Clemons, *Anal. Chem.*, **38**; 753 (1966). Turk et al., Paper, 149th Annual Meeting, American Chemical Society, Detroit, 1965. Tabor, *JAPCA*, **15**; 415 (1965).

Table 11 (cond't)

Pollutant	Sampling Apparatus	Sampling Rate (cfm)	Minimum Sample Volume (ft²)
Peroxy acyl nitrate (PAN)	100-ml gas collection bottle or gas pipette	Batch sample	0.001
	Wet impinger	0.1	1–5
Phenols	High volume sampler	40	24-hr sample
	Wet impinger, aq sodium hydroxide	1.0	0.5

cumstantially related to episodes of elevated SO_2, NO_2, and suspended particulate levels in urban industrial areas.

Furthermore, extensive interrelationships involving a large number of variables are not only possible but, in the recent work of Larson,[22] a statistical reality. Where great difficulty arises in attempting to compare and reduce various air quality data directly, Larson has developed an indirect mathematical "model" that is used to assemble such relationships, assuming that all air concentrations of pollutants have the following characteristics:

1. Pollutant concentrations are log normally distributed over all averaging times.
2. Median pollutant concentrations are proportional to averaging times raised to some exponent.
3. Maximum pollutant concentrations are inversely proportional to averaging times raised to some exponent.

Larson's equations, based on these assumed characteristics, can be used to develop predicted geometric mean, standard geometric deviation, maximum

Analytical Method	Sensitivity (ppm)	Chapter Reference	Other References
Gas chromatography electron capture detection	0.001	3	*Methods of Air Sampling and Analysis*, Intersociety Method No. 44301-01-70T (1970).
Colorimetry	0.1		Stevens and Price, "Colorimetric Analysis of PAN," Eighth Conference on Methods in Air Pollution and Industrial Hygiene Studies, Oakland, Calif. (1967).
Colorimetric	0.0013	2	Smith, MacEwen, Barrow, *AIHAJ*, April, 142-8 (1959). *Methods of Air Sampling and Analysis*, Intersociety, Method No. 17320-01-70T (1972).

once-per-year expected concentration, and expected frequency distribution of any given pollutant concentration.

A detailed treatment of this model is beyond the scope of this book and not entirely within the realm of chemical analysis methods per se. However, for the reader who is engaged in evolving community air standards and in community air surveying, a background in mathematical modeling techniques is essential.

These systems are most valuable because they deal with real data and produce usable extrapolations. They are most hazardous when they are extended to conditions that are not covered by the assumptions. Thus indiscriminate substitution of data from one region into another and inclusion of too few data, or for some reason "suspect" data, must be avoided.

Statistical Evaluation of Data

The end product of a number of air samples of a certain kind, for a certain contaminant, consists of data. Interpretation of these data requires determination

Table 12 Concentrations of Common Air Pollutants by City

City	SO$_2$ (ppm)	CO (ppm)	NO (ppm)	NO$_2$ (ppm)	Total Hydrocarbons (ppm)	Total Oxidants (ppm)
Chicago	0.125	8.4	0.076	0.050	3.00	0.028
Cincinnati	0.021	5.6	0.032	0.028	2.50	0.031
Denver	0.005	7.6	0.037	0.037	2.44	0.027
Detroit	0.029	5.2	—	0.010	2.45	0.014
New Orleans	—	—	—	—	—	—
Philadelphia	0.098	6.3	0.063	0.043	2.42	0.026
San Francisco	—	—	—	—	—	—
St. Louis	0.029	5.6	0.035	0.024	—	0.035
Washington, D.C.	0.048	4.9	0.047	0.043	—	0.025

[a] Maximum concentration 24-hr bubbler.
[b] Maximum concentration.

of (1) average conditions, (2) most predominant tendency, (3) variance between results (deviation), and (4) degree of confidence (measure of probable error). The median is often used as a measure of the central tendency, especially when the scatter of the results is so great that groups containing a few pieces of data cannot be found. Such a situation is often the case with air sampling results. For such ungrouped data, the median can be computed simply by arranging the data in order of increasing value and selecting the middle term by inspection. Together,

Table 13 Concentrations of Suspended Particulate in Common Contaminant Metals (Arithmetic Mean, NASN, 1971)[20]

City	Cadmium (μg/m^3)	Chromium (μg/m^3)	Iron (μg/m^3)	Lead (μg/m^3)	Nickel	Zinc
Burbank, Calif.	0.004	0.028	2.5	2.8	0.034	0.10
Fort Worth, Tex.	0.005	0.003	1.1	1.1	0.004	0.01
Glassboro, N. J.	0.004	0.003	0.6	0.3	0.015	0.25
Glencove, N. Y.	0.004	0.010	—	—	0.028	—
Kansas City, Kan.	0.002	0.006	1.4	0.2	0.009	0.28
St. Paul, Minn.	0.001	0.009	1.5	0.4	0.012	0.03
Salt Lake City, Utah	0.001	0.002	1.1	0.3	0.007	0.03

(Arithmetic Mean, NASN, 1971)[20]

Benzene Soluble Organics ($\mu g/m^3$)	Aldehydes[a] ($\mu g/m^3$)	Ammonium ($\mu g/m^3$)	Nitrates	Sulfates	Fluoride[b] ($\mu g/m^3$)	Suspended Particulate
5.7	25	1.6	1.6	12.0	0.42	143
5.8	21	1.1	1.9	12.7	0.16	107
8.5	21	0.1	1.7	3.7	0.23	106
6.8	54	0.7	3.3	12.0	0.57	141
6.6	—	0.2	2.7	7.1	0.50	81
11.1	27	3.4	2.0	20.6	0.52	162
6.6	—	0.6	2.8	5.6	—	75
6.3	92	1.2	1.6	14.2	0.19	124
5.4	11	1.7	1.9	11.1	—	106

the simple mean (arithmetic average) and the median of the data are the two most common expressions employed to describe the average "meaning" of a set of experimental results. However, it can easily be imagined that either of these values can be misleading in cases where values (data) are not closely clustered. For example, the typical hourly SO_2 data in Table 14 show a calculated mean that never occurred as an actual measurement.

In reality, the community that experienced these levels over the hypothetical 5-hr sampling period was not exposed to the average effect. Instead, it experienced a sudden high-level episode of exposure that began sometime between 4:00 and 5:00 P.M. For this occurrence, the "average level" has no physical meaning.

A certain degree of variance or "scatter" is also evident in Table 14, where the simplest measurement of this variance is the range. The range is the difference between the highest and lowest two measured values in the population of data. Although simple, range can be misleading, especially if the maximum or minimum value is unusually large or small.

Of wider application than the range is the standard deviation. This unit is defined as the square root of the average of all the squared deviations taken about the average. Expressed mathematically, the standard deviation is

$$S = \sqrt{\frac{\sum (X_i - \bar{X})^2 n}{N}}$$

where $(X_i - X)$, which is the difference between a piece of individual data (X_i)

Table 14 Example of Hourly Average SO₂ Concentration

Clock Time (P.M.)	SO₂ (ppm)
1:00–2:00	0.02
2:00–3:00	0.02
3:00–4:00	0.01
4:00–5:00	0.12
5:00–6:00	0.18
Mean = 0.07	

and the average of the data (X), occurs n times, and N is the total number of data entries.

Using normally occurring CO data as an example, Table 15 shows calculation of the standard deviation expressing the degree of variance over 108 hr of sampling. First determine the mean (X) by $X = \sum nx/N = 3.5$ ppm. Then, using X, determine the individual deviations $(X_i - X)$ and the $(X_i - X)^2 n$ for each value. The standard deviation for these hourly averages follows from

$$S = \sqrt{\frac{\sum (X_i - X)^2 n}{N}} = \frac{119}{108} = 1.10 \text{ ppm}$$

It is well to note here that the standard deviation is affected by the value of every entry of variance in the range. Since each variance is squared, extremely diverse

Table 15 Example of Typical CO Data (Hourly Averages)

Hours of Average CO [ppm (X_i)]	Number of Occurrences, of X (n)	nX_i	$X_i - \bar{X}$	$(X_i - \bar{X})^2$	$X_i - \bar{X}^2 n$
1.5	8	12.0	−2	4	32
2.5	26	65.0	−1	1	26
3.5	40	140.0	0	0	0
4.5	25	112.5	1	1	25
5.5	9	49.5	2	4	36
	$N = 108$	$\sum nX = 379.0$			$\sum(X_i - X)^2 n = 119$

entries affect the absolute value of S to a greater extent than do entries close to the mean. Thus a small value for the standard deviation indicates a close clustering of data around the mean, while a large S indicates rather wide scattering.

A more quantitative means of expressing this degree of scatter or variance lies in calculation of the standard error (S_x). In air sampling, this quantity represents the degree to which the scatter of data about the mean approaches normal distribution. This measure of statistical error in sampling depends on the fact that 99.7% of the cases of a normal distribution lie within three standard deviation units from the mean.

As an example, let us assume that simultaneous calibration using 10 side-by-side hydrocarbon analyzers produced an hourly average (mean) hydrocarbon concentration of 1.4 ppm, with a standard deviation of 0.40 ppm. If one of the 10 samples showed a mean for 6 replicate hours of 1.5 ppm hydrocarbon, should this result be considered in error?

Here, the population mean (X) is 1.4 ppm, and the standard deviation (S) is 0.40 for a sample size (N) of 6, which has a mean (μ) of 1.5. The standard error of the mean (S_x) in this case is

$$S_x = \frac{S}{\sqrt{N}} \quad \frac{0.40}{\sqrt{6}} = 0.16$$

Now, we assume the mean of the sample to be normally distributed and introduce Z as the standard normal variation of such a sample population of mean 1.4, where

$$Z = \frac{\overline{X}(\text{sample}) - \overline{X}(\text{population})}{S_x} \quad \text{or} \quad \frac{X - \mu}{S_x}$$

Then, on substitution

$$Z = \frac{1.5 - 1.4}{0.16} = 0.625$$

Thus the standard error of 0.16 is small when compared with three standard deviations, where the sample result in question differs from the population mean by 0.625 standard deviations. Here, we would find that such a deviation from normal distribution corresponds to a probability of 0.53. This means that in approximately 53 samples per 100, a deviation as large as this or larger would be expected. (*Note.* Most books dealing with probability include a table that relates deviation to probability of occurrence for a standard distribution.)

On the other hand, suppose the sample did not possess normal distribution with respect to the mean. What can be said about retaining the sample? A rule of thumb is that a piece of data should be held suspect or rejected if its deviation is greater than three standard deviations from the mean.

Principles of Air Sampling

If the sample is suspect, but included, what can be deduced about the distribution of the samples about the mean?

Here, we follow the theorem that states: If a variable possesses a distribution with mean ($\bar{X}$) and a standard error ($S/\sqrt{N}$), then the mean based on the sample size (n) will possess an approximately normal distribution with the mean (μ), the approximation becoming better as the value of $S/\sqrt{N}$ increases.

From this, we can see that for cases such as air sampling, where many variables enter into the sampling data, an individual average value is not required to follow normal distribution for the data to be normally distributed, although most distributions will be normal for numbers of samples greater than 10. Stating this behavior in another way, if we examine the relationship $S/\sqrt{N}$, we see that a greater absolute standard deviation is better distributed over a larger number of samples.

Sampling Times

When samples are to be collected on a random schedule by batch methods, the chemist is often faced with the question of how many samples to collect to achieve representative sampling. Since air sampling is often based on determination of the approach of air quality to a given "standard," it is useful to determine the number of samples required for confidence in the program as having a representative sampling.

If we consider p as the probability of measuring a given value X in a single event, q as the probability of measuring some other value, and X/n as the proportion of success, then it can be demonstrated[23] that the expected frequencies for $0, 1, 2 \ldots n$ successes in N trials have a mean value

$$\mu = np$$

and a standard deviation of

$$S = \sqrt{npq}$$

Substituting these values into the equation for Z, we get

$$Z = \frac{X - \mu}{S} = \frac{X - np}{\sqrt{npq}}$$

If we use the proportion of success (X/n), this equation becomes

$$Z = \frac{X/n - p}{\sqrt{pq/n}}$$

where $\mu = p$, and $s = \sqrt{pq/n}$.

For a given situation, if a given ambient air quality standard states that ambient air shall not show a measured level of 0.5 ppm oxidant measured as ozone for more than 1 hr/100 hr or 1% of the time, what is the probability that random 1-hr samples will measure air of a concentration 0.5 ppm O_3 or greater if the sample air does in fact contain O_3 in excess of 0.5 ppm for as much as 10% of the time?

Here, by substitution

$$\mu = p = 0.10$$

$$S = \sqrt{\frac{pq}{n}} = \sqrt{\frac{(0.10)(0.90)}{50}} = 0.042$$

where a probability of 10% success assumes a probability of 90% failure.

Thus

$$Z = \frac{x/n - p}{S} = \frac{0.01 - 0.10}{0.042} = -2.1$$

If a normal distribution of data is assumed; Z corresponds to a probability of detection greater than 95% for 50 samples.

SUMMARY

When planning an air pollution survey, it is important at the outset to decide whether a short-term measurement or a long-term sampling program is to be employed and at the conclusion to be certain that the expression of data is in keeping with the sampling period. Thus a sample value averaged over 1-hr sampling time can be properly compared only to a 1-hr air quality standard and should not be interpreted as an annual average, thereby inferring that air quality is normally the particular concentration measured.

Recently published United States ambient air quality standards are shown in Table 16.

The primary standard is intended for protection of population from the standpoint of health, while the secondary standard is designed to protect against damage to vegetation and materials and to ensure a sense of environmental well-being.

It should be remembered that, although short-term or "spot" sampling may demonstrate the presence of a source of a particular pollutant in a community, cycles of emission are not revealed. Continuous sampling at fixed locations is the approach of choice when the goals of the study are:

1. Evaluation of general population exposure.
2. Estimation of degree of improvement achieved by abatement measures.
3. Correlations between levels of pollutants and meteorological factors.

Table 16 Air Quality Standards, 1972

Gases	Primary Standard	Secondary Standard
SO_2 ($\mu g/m^3$) Annual arithmetic average	80 (0.03 ppm)	
CO ($\mu g/m^3$) Maximum 8-hr concentration	10 (9 ppm)	10 (9 ppm)
Maximum 1-hr concentration	40 (35 ppm)	40 (35 ppm)
Photochemical oxidants ($\mu g/m^3$) Maximum 1-hr concentration	160 (0.08 ppm)	160 (0.08 ppm)
Nitrogen oxides ($\mu g/m^3$) Annual arithmetic average Maximum 24-hr concentration	100 (0.05 ppm)	100
Non-methane hydrocarbons ($\mu g/m^3$) Maximum 3-hr avg. (6 to 9 A.M.)	160 (0.24 ppm)	160 (0.24 ppm)
Particulates		
Suspended particulates ($\mu g/m^3$)		
Annual geometric mean	75	60
Maximum 24-hr concentration	260	150

In assembling a sampling study, equipment needs should be anticipated. A regular program of calibrating air sampling equipment on a monthly or quarterly basis is recommended.

Actual extent of sampling depends on a number of practical considerations. These include:

1. The identity of the pollutant or pollutants to be measured and the approximate concentration(s), together with general information concerning the location of the sampling site that might give rise to interferences.

2. Duration of sampling desirable, and possible time limitations imposed by availability of personnel.

3. Access to ambient air to be sampled through a sample line arrangement of sampling equipment, auxiliary equipment (e.g., tables and distilled water), locating of electrical power sources.

4. Cost.

REFERENCES

1. R. I. Larson, "A Mathematical Model for Relating Air Quality Measurements to Air Quality Standards," USEPA, NERC, Office of Air Programs, Publication No. AP-89, 1971.
2. *Federal Register*, **36,** 1511 (1971).
3. Air Quality Criteria for Nitrogen Oxides, Environmental Protection Agency, Air Pollution Control Office, Wash., D.C., 1971.
4. S. A. Roach, "A More Rational Basis for Air Sampling Programs," *Am. Ind. Hyg. Assoc. J.*, **27,** 1–12 (1966).
5. B. E. Saltzman, "Significance of Sampling Time in Air Monitoring," *J. Air Poll. Control, Assoc.*, **20,** 660–665 (1970).
6. Standard Methods for Measurement of Fuel Samples, *ASTM,* **19,** 154, D 1071-55, revised 1970.
7. H. B. Weiser, *Colloid Chemistry*, Wiley, New York, 1958, pp. 49–54.
8. "Recommended Standard Methods for Continuing Dustfall Surveys," (APMI, Revision 1) Environmental Protection Agency, NAPCO, *J. Air Poll. Control Assoc.*, **16,** 372–376 (1966).
9. American Society for Testing Materials, ASTM, Standard Method D-1739-62, 1962.
10. J. S, Nader, "Dust Retention Efficiencies of Dustfall Collectors," *J. Air Poll. Control Assoc.*, **8,** 35–39 (1958).
11. H. P. Sanderson, P. Brandt, and M. Katz, "A Study of Dustfall on the Basis of Replicated Latin Square Arrangements of Various Types of Collectors," *J. Air Poll. Control Assoc.*, **8,** 461–466 (1963).
12. J. Stockham, S. Radner, and E. Grove, "The Variability of Dustfall Analysis Due to the Container and Collecting Fluid," *J. Air Poll. Control Assoc.*, **16,** 263–267 (1966).
13. D. H. Lucas, "Certain Aspects of the Deposition of Dust," *J. Inst. Fuel,* **30,** 623–625 (1957).
14. *Federal Register,* **36**(84), 8191–8193 (1971).
15. H. C. McKee, R. E. Childers, and O. Saenz, Jr., "Collaborative Study of Reference Method for the Determination of Suspended Particulate in the Atmosphere (High Volume Method)," Southwest Research Institute, Houston, Texas, Contract CPA 70-40, June 1971.
16. J. C. Parke, "Developments in the Use of the A.I.S.I. Automatic Smoke Sampler," *J. Air Poll. Control Assoc.*, **10,** 303–306 (1960).
17. E. Kemeney, "The Determination of Gravimetric Pollution Concentrations by Means of Filter Paper," *J. Air Poll. Control Assoc.*, **12,** 278–281 (1962).
18. W. T. Ingram, "Smoke Curve Calibration," paper presented at the 1971 Meeting of the Air Pollution Control Association, Atlantic City, N.J., paper No. 71-79.

19. A. C. Stern, ed., *Air Pollution,* Vol. II, Academic Press, New York, 1968, p. 33.
20. *Air Quality Data for 1967* (Revised 1971), National Air Surveillance Network, Division of Atmospheric Surveillance, Environmental Protection Agency, August 1971.
21. G. R. Hilst and J. G. Bryan, Preliminary Meteorological Analysis of National Air Sampling Data," I, II, Contract Cwb-10014, Travelers Res. Center Inc., Hartford, Connecticut, 1962.
22. R. I. Larson, *A Mathematical Model for Relating Air Quality Measurements of Air Quality Standards,* Environmental Protection Agency, Office of Air Programs, November 1971.
23. H. L. Alden and E. B. Roessler, *Probability and Statistics,* W. H. Freeman Co., San Francisco, 1968, p. 89.

6

CALIBRATION OF SAMPLING INSTRUMENTS AND PREPARATION OF STANDARD GAS MIXTURES

INTRODUCTION

To ensure reliable air sampling data, the air sampler and the analytical method must be calibrated using a sample of air or a synthetic gas mixture in which the concentration of the contaminant is measured to a degree of accuracy well within the limit of sensitivity of the analytical method.

The method of preparing standard contaminant atmospheres and of calibrating air sampling equipment often has been included with the method of analysis as presented in this text for the particular contaminant. However, because of the importance of this step in ambient air sampling and analysis, this separate discussion will acquaint the reader with the general approach to the preparation and use of calibration mixtures.

APPROACHES TO CALIBRATION

The mode of operation of the air sampler selected usually determines the approach required for its calibration. For example, a continuous sampler that consumes 160 ml/min of sample air and requires a 10 to 15 min response time must be furnished with at least 1600 ml of calibration mixture prior to the appearance of any usable calibration data. Consequently, to ensure such a sampler a minimum 1-hr calibration cycle, slightly more than $\frac{1}{3}$ ft^3 of standard gas mixture is required. Here, a rule of thumb in preparing such a calibration mixture is to prepare from 2 to 10 times the volume of gas mixture required by calculation, so that a sufficient quantity is available for rerun of the calibration cycle or for subsequent check or comparison to other analysis methods. However, it is not good practice to attempt to retain a static volume of calibration gas for a period of time longer than a few hours to a few days. If the gas is stored for longer periods, such effects as adsorption, sunlight activity, and permeation through the container walls may change the analytical equivalence of the mixture.

Preparation of calibration gases requires careful attention to (1) the purity of the standard contaminant to be diluted, (2) the purity of the dilution gas (usually air), (3) the inertness of storage containers, and (4) considerations involving sample gas transfer systems.

Purity of Materials Used to Prepare Standard Gas Mixtures

Chemically pure reference gases can be purchased either as compressed gases, as liquids, or, in some cases (e.g., CO_2), as weighable solids. Particulates may often be secured directly from the process involved and used as reference materials. For pure vaporized or sublimed metals or salts, chemically pure reagents should be purchased.

Permeation tubes (i.e., semipermeable cylinders containing pure reference contaminants in liquid form) are commercially available for most normal contaminant gases. Here, weight loss of the tube is used as a measure of the contaminant added to the diluent gas.

Probably the most convenient reference or calibration gases are commercially prepared reference mixtures whose composition is certified by analysis (usually chromatographic).

Purity of the Dilution Gas

Standard dry compressed cylinder air or nitrogen are the most common diluent or "vehicle" gases. However, impurities of several types may be present in trace amounts prior to compression or may be introduced during compression of the gases. While the most common of such contaminants are water vapor and oil mist, other contaminants often present are CO_2, CO, acid gases, and even dust particles resulting from such sources as pipe corrosion. These contaminants must, of course, be removed prior to use of the gases.

Separating Water Vapor

Moisture is generally removed, prior to adding the reference component, by desiccation, cooling, or a combination of both. Here, ordinary laboratory air may often be sufficiently purified to use as a diluting medium for a contaminant that is known to be absent from the laboratory air simply by passing the laboratory room air through a desiccant such as Drierite, followed by Ascarite (to remove CO_2) and a dry ice trap in this order.

An appropriate desiccant can be selected from a number of solids on the basis of its performance, convenience of handling, or other considerations. Although *magnesium perchlorate* has the greatest efficiency and highest capacity of any solid for absorbing moisture, its high oxidizing potential requires that it be protected from contact with air containing entrained organic vapors with which it may form explosive mixtures. Calcium sulfate (Drierite or Anhydrocel) is probably the most common desiccant, and it is easily regenerated by heating for 1 to 2 hr at 200°C. Other less frequently employed desiccants are silica, alumina, and calcium chloride.

Separation of Oil Mists and Organic Compounds

Moderate molecular weight ($nC > 2$) hydrocarbons are generally separated from air by passing the air through a drying tube that has been packed with 4 to 20 mesh coconut charcoal. Alternatively, a small combustion tube packed with copper turnings and heated to 675°C is used to convert even low-molecular-weight hydrocarbons to CO and CO_2. After cooling the effluent gas, CO and CO_2 are removed by scrubbing with chromic acid and filtering through Ascarite, followed by charcoal and finally a plug of glass wool.[1]

Separation of Particulates and Other Air Contaminants

A number of inert filter media, fabricated of steel mesh, metal fibers, and porous sintered sieve is available to separate particulate to a lower size limit of 0.1 μ. These and other metallic filters possess the mechanical strength to resist rupture at flows up to 1 ft^3/(min) (s $in.^2$) surface. Such filters can remove water, heavy hydrocarbons, and solid particulate. Chemically reactive materials, such as acid gases, are removed by a 4 to 20 mesh packing of soda lime, while CO_2 is adsorbed by Ascarite. Hopcalite, described in Chapters 3 and 4, is used to convert CO to CO_2, which is subsequently adsorbed by Ascarite.[2]

APPARATUS REQUIRED TO PREPARE CALIBRATION GAS MIXTURES

Storage Containers

Containers used for the storage of calibration gases can be divided into rigid and flexible vessels. These are usually required when a "batch" or static volume of calibration gas is to be prepared.

Rigid containers are usually fabricated of either soda or borosilicate glass. However, polyolefin bottles have recently been employed with little observed surface reactivity.[3] Stainless steel and Teflon are often employed as chamber materials in spite of their rather high cost.

Ordinarily, such vessels are bottles of between 20 and 60 liters (5 to 15 gal) capacity. They are fitted with a stopper, screw cap, or lid that can accommodate an inlet and outlet pipe, an impeller or plunger that acts as a mixer, and a rubber septum through which a precise volume of calibrant is injected by means of either a liquid or gas-tight syringe or a micropipette.

The volume of these containers is determined simply by filling with water or some weighable liquid of known density.

Flexible containers offer the advantage of greater economy and portability over the usually heavier rigid storage bottles. Such calibration "bags" can be filled in the laboratory and then easily transferred to the field for on-site calibration of an instrument. In addition, upon respiration (i.e., adding or subtracting calibration gas), a negligible change in pressure occurs. Thus makeup air or partial volume correction is not needed in the course of instrument calibration. Bags are usually fabricated of Teflon, polyolefin, vinyl, or vinyl chloride material or plastic-coated aluminum or other flexible metallic fiber. Desirable properties of such plastics are chemical inertness, impermeability, and resistance to cracking or temperature change. One of the most satisfactory materials is Mylar, which is a polyester.

Other variable-volume containers include (1) piston-type cylinders, in which diluent gas expands against a movable piston to a volume determined by a calibrated scale along the traverse of the piston, and (2) bladders that, when inflated, assume a final volume imposed by the walls of an outer rigid chamber.

Transfer Systems

Volumetric equipment that is employed in transferring reference compounds and their subsequent dilutions is generally fabricated of glass, with moving parts (e.g., plungers, valves, or mixing apparatus) made of Teflon, ground glass, or butt-to-butt glass tubing sleeved with flexible polyvinyl or polyolefin elastomer.

METHODS OF CALIBRATION

When a limited amount of laboratory equipment is available, simple standard laboratory glassware can be employed to produce a static volume of calibration gas. This approach is known as the single volume or static method for instrument or bubbler calibration. Alternatively, a number of advantages may be gained by continuous preparation of the calibration mixture. The dynamic method of generating gas or vapor mixtures, which requires somewhat more specialized equipment, yields an uninterrupted flow of calibration gas that is less subject to error due to in-transit decomposition or adsorption on the walls of the container.

Batch Method of Preparing Calibration Gas

The addition of a weighed or otherwise measured amount of pure contaminant into a large (e.g., 5 to 10 gal) glass jug with adequate vaporization and mixing constitutes the basis for preparation of a single batch volume of calibration gas.

Contaminants (e.g., aldehydes or mercaptans) that are ordinarily liquids at atmospheric pressure can be introduced by using a microsyringe to inject them directly into a container holding a specified captive volume of purified air, allowing from 12 to 48 hr to vapor-equilibrate. Alternatively, the liquid can be injected by means of a stream of a pure air, which is used to fill a clean evacuated jug until atmospheric equilibrium is established. The jug of known or calculated volume then contains a measured amount of contaminant that can be converted into micrograms per cubic meter or parts per million.

Procedure for Laboratory Calibration Gas

Ordinarily, such a jug is assembled as shown in Figure 1 and filled in the following manner.

1. Purge the container jug with clean, dry air by drawing a vacuum at A, leaving stopcock C open to room air or other "clean" air (a drying tube may be attached at C to dry room air). Close stopcock C. After a 1-min pause, close stopcock A, leaving the jug under slight negative pressure.

2. Introduce a known amount of pure contaminant at B, preferably by injection through a septum, with the temperature of the hot plate set above the boiling

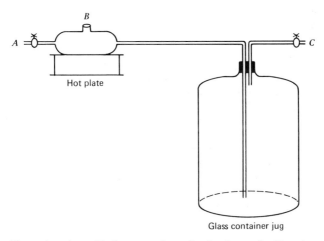

Figure 1. Assembly for preparing a fixed volume of calibration.

point of the contaminant. Immediately open stopcock A slightly to allow clean dry air to sweep the contaminant vapor into the jug.

3. Mix the gases by means of a finned Teflon stirring bar, Teflon "confetti," or Teflon slugs through successive careful inversions of the jug. Gas phase solutions also may be mixed by alternate applications of cold and heat to opposite extremities of the container.

Small amounts of this calibration gas can be withdrawn at the septum by means of a gas-tight syringe, without regard for dilution effect. Alternatively, a stream of the mixture can be withdrawn provided several equivalent bottles, each filled with an equal amount of calibration gas, are arranged in a series. Thus when larger amounts of calibration gas are removed from the "prime" bottle, makeup air will be composed essentially of calibration mixture, thereby resulting in a very small change in concentration of the prime container.

It is noteworthy that, as an alternate procedure, contaminant gas contained in a fragile ampule can be placed in the static bottle, the bottle stoppered, and the ampule broken by a sharp twist or shake of the bottle.

Finally, a gas burette that has been purged to equilibrium and filled with contaminant from a compressed gas cylinder can be flushed into the glass dilution container with clean dry air.

Procedure for Field Calibration Gas

A more convenient method for field calibration avoids the use of a glass jug by employing flexible plastic bags or bladders. These containers are filled in the following manner:

1. The bag is first flushed to remove residual contents by repeated filling and deflating with clean dry air. It is then evacuated to slight negative pressure and folded or rolled to ensure escape of any entrapped air.

2. A length of Teflon tubing, attached to the bag face by means of a valve, is then connected through a rotameter or dry gas meter to a source of clean dry air.

3. A known volume of contaminant gas is then introduced into one of the following:

    a. The flowing stream of diluting air.
    b. A heated reservoir upstream of the bag.
    c. The bag directly by injection through a soft rubber patch.

The volume of dilution air introduced is measured by the standard flow device.

4. Thorough mixing of the gas mixture is achieved through manual flexing of the bag.

As calibration gas is withdrawn, the walls of the bag conveniently collapse so that no volume of dilution air need be accounted for.

Other static containers, such as those equipped with a movable piston or a bladder that remains at atmospheric pressure, inside a rigid container, are described in several other references.[4-6]

Calculations

Assuming 100% pure standard contaminant, its volume concentration in pure dilution air can be calculated quite simply, whether the contaminant is added as a gas or liquid.

When a gas-tight syringe is filled by withdrawing pure sample gas from a plugged sleeve or tubing attached to a gas cylinder, the resulting volume percent concentration at atmospheric pressure is

$$\frac{C_g}{V} = \frac{V_s \times 100}{V_c}$$

where C_g is the desired volume percent concentration to be used as calibration gas, V_s is the volume of sample withdrawn in milliliters, and V_c is the measured volume of the container in milliliters, or the volume of clean dry diluent gas (in milliliters) delivered through a rotameter to a flexible bag.

Solving the above equation for V_s allows calculation of the sample volume needed to produce some desired concentration C_g.

Injection of a liquid requires conversion to the vapor volume to determine the volume percent, C_l.

Here,

$$C_l = \frac{V_s}{V_{ec} \times MW} \times \frac{d_1 \times T \times 760 \times 22{,}400 \times 10^2}{273.1 \times P}$$

where MW is the molecular weight of the contaminant, P is the atmospheric pressure in millimeters of Hg, d_1 is the density of the liquid contaminant, and T is the temperature in degrees Kelvin.

When it is desirable to express the concentration in parts per million, the concentration can be calculated by:

$$C_{\text{ppm}_g} = \frac{V_s \times 10^6}{V_c}$$

and

$$C_{\text{ppm}_l} = \frac{V_s}{V_c \times MW} \times \frac{d_1 \times T \times 760 \times 22{,}400 \times 10^6}{273.1 \times P}$$

In cases where a very small capacity container (V_c) is used to inject as little as 1 to 50 μl of liquid (rather than gaseous) contaminant, the liquid does not entirely wet the delivery volume of the syringe. Therefore, a slight correction should be made for this empty delivery volume by calibrating the syringe, using the material to be injected, and either weighing the delivered volumes or, for volumes less than 0.1 ml, using spectroscopy to establish plot of apparent volume versus measured volume.[7]

As calibration gas is removed from a storage bottle, with subsequent replacement of the volume withdrawn by clean, dry dilution air, a correction must be made for the dilution volume. Also, the volume withdrawn, together with the time or date, should be entered on a log attached to the wall of the bottle. Each new successive concentration in the storage bottle is calculated from the prior concentration by

$$C_2 = C_1 \times e^{-V_r/V_c}$$

where C_1 and C_2 are the initial and final concentrations, V_r is the volume of calibration air removed (or the volume of replacement air as pure diluent gas), and V_c is the volume of the container. Obviously, this calculation is not required in dealing with flexible bags as containers.

As described previously, this dilution effect can be minimized and the strength of the calibration mixture maintained near the desired concentration by placing several large jugs containing calibration gas at the same concentration in series. Here,

$$C_2 = C_1 \left[1 + \frac{1}{\lfloor 1} \left(\frac{V_r}{V_c}\right) + \frac{1}{\lfloor 2} \left(\frac{V_r}{V_c}\right)^2 + \frac{1}{\lfloor n-1} \left(\frac{V_r}{V_c}\right)^{n-1} \right] \cdot e^{-V_r/V_c}$$

where C_2 is the effective concentration, after removal of volume V_r, and C_1 is the initial concentration of each jug of volume V_c. Note from the last term, that the number of terms employed is one less than the number of bottles.

Example Problems

1. What concentration in volume parts per million of methylene chloride (CH_2Cl_2) results from the injection of 0.5 ml liquid into a partially evacuated 34.8-liter glass container at 20°C with subsequent dilution using clean dry laboratory air to an ambient pressure of 760 mm Hg?

$$C_{ppm} = \frac{0.5 \text{ ml}}{34{,}800 \text{ ml} \times 85 \text{ g/mole}} \times \frac{1.34 \times 293 \times 760 \times 22{,}400 \times 10^6}{273.1 \times 760}$$

$$C_{ppm} = 0.5 \text{ ppm}$$

2. What volume of SO_2 gas must be injected into a 10-gal glass container so that a concentration of 150 ppm is achieved when dilution air is added to reach ambient pressure of 760 mm Hg at 25°C?

$$V_c = 10 \text{ gal} \times 1 \text{ ft}^3/7.5 \text{ gal} \times 28.3 \text{ l/ft}^3 = 37.7 \text{ liters}$$

Substituting in the equation

$$V_s = C_{ppm} \times V_c \times 10^{-6}$$
$$V_s = 150 \times 283{,}000 \text{ ml} \times 10^{-6}$$
$$V_s = 5.65 \text{ ml}$$

Since the volume of the injected gas is subject to ambient temperature and pressure, these terms are deleted from the volume parts per million calculation.

3. What concentration of CO remains in a 10-liter container after 2 liters of calibration gas have been removed, if the original concentration of the gas was 300 ppm?

$$C_2 = C_1 e^{-V_r/V_c} = 300 \text{ ppm} \cdot e^{-2/10}$$

Taking the log of both sides of the equation,

$$2.3 \log \frac{C_1}{C_2} = \frac{V_r}{V_c}$$

Then,

$$2.3(\log 300 - \log C_2) = 2/10$$
$$\log C_2 = 2.46$$
$$C_2 = 288 \text{ ppm}$$

4. Exactly how many minutes must a dilution airflow of 12 liter/min be maintained to establish a concentration of 0.06 ppm SO_2 in a Mylar bag if the initial injection of SO_2 was 7.2 μl gaseous SO_2?

$$V_c = \frac{V_g \times 10^6}{C_g} = \frac{7.2 \times 10^{-3} \text{ ml}}{0.06 \text{ ppm}} \times 10^6 = 120{,}000 \text{ ml} = 120 \text{ l}$$

Therefore,

$$t = \frac{120 \text{ l}}{12 \text{ l/min}} = 10 \text{ min}$$

5. If three 10-liter bottles, each containing 0.100 ppm calibration gas, are placed in series and instantaneous mixing is assumed, what is the concentration of NO_2 after intervals of 10 min, ½ hr, and 2 hr if the analyzer requires a flow of 100 ml/min?

$$V_r (10 \text{ min}) = 10 \text{ min} \times 100 \text{ ml/min} = 1 \text{ liter}$$
$$V_r (30 \text{ min}) = 30 \text{ min} \times 100 \text{ ml/min} = 3 \text{ liters}$$
$$V_r (2 \text{ hr}) = 120 \text{ min} \times 100 \text{ ml/min} = 12 \text{ liters}$$

$$C_{(10\text{ min})} = 0.100 \text{ ppm} \left[1 + \frac{1}{\underline{1}}\left(\frac{1}{40}\right) + \frac{1}{\underline{2}}\left(\frac{1}{40}\right)^2\right] e^{-1/40} = 0.090 \text{ ppm}$$

$$C_{(30\text{ min})} = 0.100 \text{ ppm} \left[1 + \frac{1}{\underline{1}}\left(\frac{3}{40}\right) + \frac{1}{\underline{2}}\left(\frac{3}{40}\right)^2\right] e^{-3/40} = 0.075 \text{ ppm}$$

Similarly,

$$C_{(2\text{ hrs})} = 0.035 \text{ ppm}$$

Precautions

Consideration should be given to use of a greaseless transfer system (e.g., Teflon stopcocks and joint sleeves), since the stopcock lubricant of ground-glass systems can absorb small amounts of trace contaminant gas during introduction into the storage container. Also, the use of tubing other than glass or Teflon in injection reservoir or transfer materials should be avoided.

Purity of both the standard contaminant and dilution gas should be verified, and a system of purifying and drying filters as described in the section Purity of the Dilution Gas should be employed in transferring dilution gas (or laboratory air) to the storage container before contact with solute contaminant.

Dynamic Methods of Producing Calibration Gas

When continuous and uninterrupted delivery of calibration gas is required to calibrate a continuous analyzer or bubbler or when very reactive or unstable calibration gases (e.g., ozone) are employed, large volumes of test mixtures at concentrations as low as a part per billion can be produced conveniently and concentrations varied using dynamic calibration methods. Although decay of the calibration gas with time and adsorption on the container walls represent serious problems for static methods, such considerations are negligible in dynamic systems.

This approach is especially useful in calibrating continuous direct-reading air analyzers, such as those described in Chapter 4.

Continuous Dilution Method

Using as chemically inert a gas manifold and flow system as possible, one or more gases can be blended to produce a calibration mixture by simply attaching the flow regulators of pure diluent and contaminant gases to a common manifold from which the flows, as measured by individual rotameters, are admitted to a blending chamber. The calibration mixture is continuously withdrawn from this chamber. Such a system is shown in Figure 2.

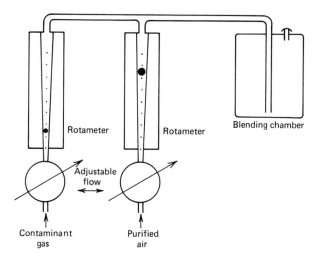

Figure 2. Flow system for continuous mixing of two or more gases to produce a calibration gas.

Adjustments in concentration of the calibration gas can be made simply by adjusting the flow of dilution air, with the calculation of the airflow-concentration ratio following directly from partial flow rates:

$$C_c = \frac{f_c}{f_d \times f_c} \times 10^6$$

where C_c is the volume concentration of contaminant gas in parts per million, and f_d and f_c are the flow rates of the diluent and contaminant gases, respectively.

Permeation Tube Methods

Nearly all plastic materials are capable of absorbing small quantities of gases. When the gas is retained in a tube or capsule of plastic at a pressure under which some of the gas is maintained as a liquid, the gas, which dissolves in the tube walls, diffuses at a calculable rate. This rate is an exponential function of the absolute temperature and a direct function of both the surface area of the tube and the thickness of the wall.

In practice, the rate of diffusion is also affected by the molecular weight of any carrier gas that comes in contact with the outside walls of the permeation tube. In addition, since some contaminant gases (e.g., SO_2 and H_2S) react with moisture, the permeation rate will be affected by any water vapor adsorbed on the tube surface, either from the atmosphere, carrier gas, or fingerprints from improper handling. Therefore, permeation tubes containing these gases should be stored in an atmosphere of dry nitrogen or, alternatively, refrigerated dry air.

Generally, polyfluorocarbon resins, such as Teflon, are sufficiently inert and most often constitute the diffusing membrane walls of these tubes. Gases that are most easily adaptable to permeation-tube dynamic calibration systems are those that are ordinarily vapors somewhat below room temperature, with critical temperatures above 20 to 30°C. Typical gases are H_2S, NO_2, SO_2, benzene, and 1-butene.

These tubes are available commercially at a cost of about $20.00 to $35.00 per tube; however, with some practice, a chemist or technician can assemble a comparable tube using ¼-in. ID Teflon tubing 1 to 10 in. in length. Liquid from a compressed gas cylinder is added to fill the tube for two-thirds of its length, with glass beads used to plug each end.[8] Permeation disks also can be fabricated in a few hours of laboratory time by cutting a 1-cm diameter disk from ¹⁄₁₆-in. Teflon stock to fit the open insert rim of a ½ × ¾ in. screw-cap stainless steel cylinder. The cylinder is two-thirds filled with liquid permeate, and the stainless steel rim is screwed down tightly over the disk and acts as the sole diffusing surface.[9]

The fundamental behavior of gases diffusing through a membrane is described by Fick's law,

$$D = dsa \left(\frac{P_1 - P_2}{W} \right)$$

where D is the volume of gas diffusing, d is the diffusion constant of the gas, s is the solubility of the gas in the membrane material, a is the membrane area, W is the thickness of the tube (or disk) wall, and P_1 and P_2 are the respective total pressures on the inside and outside walls of the membrane.

For any given permeation tube, all of the variables are fixed by the geometry of the tube except for the gas solubility (s) and the diffusion rate (d), which are exponential functions of the absolute temperature. Thus if a tube is to exhibit a constant permeation rate, great care must be taken to ensure that its temperature remains constant during the calibration procedure. Also, sufficient time must be allowed for the permeation tube to reach equilibrium in the apparatus before initiating a calibration cycle.

Ordinarily, a constant temperature, such as 30° ± 0.02°C, is chosen at which to calibrate, with the temperature regulated by an oven such as a gas chromatography oven.[10] The actual permeation rate of the tube is determined by periodic measurement of weight loss, recorded as a plot of observed weight loss versus time. A microbalance of good accuracy is usually required, since observed weight loss is often on the order of milligrams per day. Results of weight loss measurements are ordinarily expressed in nanograms per centimeter of tube length per minute. Some example permeation rates in nanograms per centimeter per minute are 1200 for NO_2, 290 for SO_2, and 250 for H_2S when the temperature is 30°C.[11]

Since permeation rates vary with respect to material and wall thickness, a range of tubes having various parts per million permeating strengths can be prepared (or purchased) in increasing lengths. Various instruments or bubblers require permeation tubes capable of supplying a representative concentration of calibrant depending on (1) the required flow and (2) the normal range of ambient concentrations encountered.

For example, when a SO_2 monitor ordinarily operates at a specified sample airflow of 100 cm³/min in an atmosphere of normal ambient SO_2 concentration equal to 0.22 ppm, the required tube length can be calculated from

$$L = \frac{R_r}{R_m}$$

where L is tube length in centimeters, R_r is the required rate of permeation in nanograms per minute, and R_m is the measured rate of a similar tube material in nanograms per centimeter per minute. First, convert parts per million to nanograms per cubic centimeter

$$\frac{\text{ppm} \times \text{MW}}{2.404} = \frac{0.22 \times 64.04}{2.404} = 59.0 \text{ ng/cm}^3$$

Then, it follows that:

$$59.0 \text{ nm/cm}^3 \times 100 \text{ cm}^3/\text{min} = 5900 \text{ ng/min}$$

Thus

$$L = \frac{5900 \text{ ng/min}}{290 \text{ ng/(cm) (min)}} = 20.4 \text{ cm}$$

After a calibrated permeation tube has been prepared, an air-sampling instrument can be easily standardized by installing the permeation tube up stream of the instrument flow duct that is placed in a constant-temperature bath or oven, as shown in Figure 3.

Preparation of a NO_2 Permeation Tube

Research in producing a satisfactory permeation tube for NO_2 has met with limited success, partly due to the saturation limit of the walls of Teflon tubes, which absorb moisture with the same rate into the tube as NO_2 diffuses out. In place of a tube, the Wayne County Air Pollution Control laboratory has experienced success in using a permeation plate prepared from a 1-cm-diameter stainless steel tube, which is filled with liquid NO_2 from a lecture bottle and then fitted with a stainless steel cap containing a ½-cm Teflon disk at its center. Such cylinders have been checked gravimetrically for in-laboratory and field reliability, with correspondence between weight loss and permeation rate.

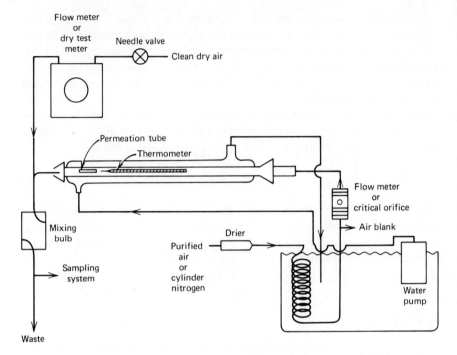

Figure 3. Laboratory system for instrument or bubbler calibration using a permeation tube at constant temperature. (Reprinted by permission from American Public Health Association, *Methods for Air Sampling and Analysis*, p. 28, 1972)

It is usually necessary to immerse the apparatus containing the permeation tube in a constant-temperature bath or thermostatic oven (20 to 30°C) that will maintain a constant known temperature to within +0.1°C. It is also advisable to protect the tubes from undue exposure to moisture through predrying of the sample air by means of an in-line Drierite filter ahead of the permeation tube.

Instrument Calibration Using a Permeation Tube

Most calibrations involve a permeation tube of fixed length (e.g., 2, 6, 10, 15 cm) that provides a permeation strength close enough to the required rate (R_r) so that only a slight variance in bubbler or instrument flow is required to achieve the desired instrument-related parts per million concentration of contaminant in the calibration airflow. The general procedure for such a calibration is described in the following section.

Procedure

After procuring or assembling a Teflon permeation tube that yields a permeation rate within the desired range, assemble the apparatus as shown in Figure 3.

A constant-temperature water bath is maintained at 20–30°C ± 0.02°C. For calibration of short duration (e.g., ½ hr), a 10 to 20 gal tank or basin, filled with water and allowed to come to room temperature for 4 to 6 hr, is an adequate substitute for a constant-temperature bath when used for immersion of the tube and condenser jacket.

A thermometer reading to 0.01°C should be used to monitor the temperature and thus prevent drift.

The diluent gas (either dry air or nitrogen) is adjusted to the temperature of the permeation chamber by prior passage through a 2-m length of copper tubing, which is immersed in the water bath. A calibrated permeation tube is placed in the outboard cooling mounted condenser and a regulated flow of purified dry carrier gas is allowed to pass through the copper tubing and across the permeation tube at a rate of 50 to 150 cm^3/min. The emergent gas stream, which passes through Teflon tubing containing permeant air mixture, is further diluted to the exact desired ppm calibration concentration by blending with dry air, with airflow adjusted by means of a flowmeter or dry gas meter to between 200 and 1000 cm^3/min. The flow rate intake of the samples to be calibrated determines the lower limit of the flow range of diluent gas.

Generally, a tube that permeates at a rate of 0.1 μl/min (0.26 μg/min) will produce a working concentration range of 0.007 to 0.4 ppm. Greater permeation rates can be achieved by using longer permeation tubes.

Schematic diagrams showing modified apparatus for in-field calibration are described in the literature.[12,13]

Preparation of Calibration Curves

Both sampling rate and sampling time control the calibration colorimetry of a fixed or "batch" sampler, such as a bubbler, while only sampling rate controls the readout for continuous sampler. Therefore, it is desirable for the purpose of illustration to present the procedure for bubbler calibration, where continuous analyzer calibration curves usually follow simply from a continuous recorder trace.

In bubbler absorption of a gas, followed by color development and measurement at a specific wavelength, the volume of sample collected determines the intensity of color produced at a fixed sampling rate. Thus a 30-liter calibration volume, which requires a flow rate of 1 liter/min for 30 min or 0.5 liter/min for 1 hr, permits rather short but practical calibration times for bubbler sampling techniques. Flow rate of a calibration gas is easily controlled by a 22-gauge hypodermic needle, which acts as a critical orifice.[14]

The flow of diluent air (f_d), which is used to adjust the concentration of the calibration mixture flow (f_c), is specified in the equation

$$(f_d + f_c) = \frac{R_p}{C \times d_v}$$

Table 1 Example of SO_2 Data for Bubbler Calibration Using Colorimetric Analysis of Absorbing Solution

SO_2 Concentration (ppm)	Measured Absorbance
0.005	0.010
0.015	0.030
0.025	0.063
0.050	0.120
0.100	0.241
0.200	0.470

where R_p is the permeation rate in nanograms per minute, C is the desired calibration concentration in parts per million, d_v is the vapor density in micrograms per microliters, and f_d and f_c are the flow rates in milliliters per minute.

Typical calibration data that result from colorimetric analysis of bubbler contents after absorption of 30 liters of calibration gas are presented in Table 1, with SO_2 used as an example.

Using the data, as above, in which the flow rate of diluent air has been adjusted to yield various part per million concentrations, plot the known parts per million SO_2 (or other calibration gas) against the absorbance of the solution in the bubbler to yield a straight line over a range of absorbance. Here, assuming no change in stoichiometry, a change in slope either at low or high concentrations indicates a variation in bubbler collection efficiency where the useful calibration range includes only data that lie on the straight line.

Example Problems

1. Given a 2.5-cm NO_2 permeation tube whose measured permeation rate is 1200 ng/(cm)(min), what single flow of dilution air is required, at 30°C and 760 mm, to yield a calibration gas of 0.10 ppm?

When a single flow rate is employed, the concentration of calibration gas (C) follows from the permeation rate, where

$$R_p = 2.5 \text{ cm} \times 1200 \text{ ng/(cm)(min)} = 3000 \text{ ng/min}$$

The parts per million concentration follows from the volume concentration of the gas at 30°C, where the calculation of vapor volume is included in the following equation:

$$C_{ppm} = \frac{R_p \times P_{(atm)} \times T/273 \times 22.4}{f_d \text{ (ml/min)} \times M}$$

Rearranging,

$$f_d = \frac{R_p \times P \times T \times 22.4}{C \times M \times 273} = \frac{3000 \times 303 \times 22.4}{0.10 \times 46.0 \times 273} = 16.21 \times 10^3 \text{ ml/min}$$

$f_d = 16.21$ liters/min

2. If the vapor density of SO_2 at 30°C and 1 atm pressure is 2.62, what flow of diluent gas must be used to dilute the 50 ml/min effluent of a permeation system to achieve a final calibration gas of 0.05 ppm, if the measured permeation rate of the tube is 290 ml/min?

$$f_d = \frac{R_p}{C \times d_v} - f_c = \frac{290}{0.05 \times 2.62} - 50$$

$f_d = 2210$ ml/min

3. What is the vapor density of H_2S at 30°C and 1 atm pressure? From the ideal gas law,

$$d_v = \frac{PM}{RT} = \frac{1 \times 34.08}{24.45} = 1.391 \text{ }\mu g/\mu l$$

REFERENCES

1. H. L. Kusnetz, B. E. Saltzman, and M. E. Lanier, "Calioration and Evaluation of Gas Detecting Tubes," *Am. Ind. Hyg. Assoc. J.*, **21**, 361–373 (1960).
2. A. B. Lamb, W. C. Bray, and J. C. W. Frazer, "The Removal of Carbon Monoxide from Air," *Ind. Eng. Chem.*, **12**, 213–221 (1920).
3. H. P. Williams and J. D. Winefordner, "Construction and Operation of a Simple Exponential Dilution Flask for Calibration of Gas Chromatographic Detectors," *J. Gas Chromatog.*, **4**, 271–272 (1966).
4. W. E. Gill, "Detection of NO_2 and NO in Air," *Am. Ind. Hyg. Assoc. J.*, **21**, 87–96 (1960).
5. G. O. Nelson, *Controlled Test Atmospheres*, Ann Arbor Science Publishers, Ann Arbor, Mich., 1971.
6. *Methods for Air Sampling and Analysis*, American Public Health Association, Washington, D.C., 1972.
7. W. H. King and G. D. Dupre, "Critique of Some Conventional Evaluation Methods and a Continuous Flow, Steady State Blender for Evaluation of Gas Chromatography," *Am. Ind. Hyg. Assoc. J.*, **28**, 79 (1968).
8. A. L. Lynch, R. F. Stalzer, and D. T. Lefferts, "Methyl and Ethyl Mercury Compounds—Recovery from Air and Analysis," *Am. Ind. Hyg. Assoc. J.*, **28**, 79 (1968).
9. L. Saad and P. O. Warner, "Preparation of Stainless Steel Permeation Disks,"

internal report, Wayne County Health Department, Air Pollution Control Division, Detroit, Mich., 1972.
10. *Methods of Air Sampling and Analysis,* Intersociety Committee Publication, American Public Health Association, Wash. D.C., 1972, p. 22.
11. *Ibid.,* p. 24.
12. *Ibid.,* p. 29.
13. *Analysis of Atmospheric Inorganics,* U.S. Dept. of HEW, EPA, NAPCO, Cincinnati, Ohio, 1967, Section VII-2, p. l.
14. J. P. Lodge, Jr., J. B. Pate, B. E. Ammons, and B. A. Swanson, "The Use of Hypodermic Needle as Critical Orifices in Air Sampling," *J. Air Poll. Control Assoc.,* **16,** 197 (1966).

7

ODOR DETECTION AND DETERMINATION

INTRODUCTION

Information is received by man through pathways called senses. Ordinarily, information passes through these senses by the transfer of energy from an object through the sense organ to the brain. For sight and hearing, the information normally travels through the air, causing vibrations that are translated by the appropriate sense organ into recognizable bits of communication. For the other three senses of touch, smell, and taste, information results from actual contact of the sensed material with the cells of the sense organ.

Light and sound lend themselves well to physical quantitation. Therefore, degrees of human exposure to either of these forms of energy can be monitored easily by electronic equipment so that harmful overexposure is avoided. Sight can be used to aid in distinguishing sense objects that may be harmful to the touch. However, ultimate determination of the harmfulness of the final two senses—taste and smell—rests with the human sensor.

Attempts to duplicate the response of the human nose date back to the first olfactometer, which was devised by Zwaardemaker in 1895. This instrument consisted of two concentric tubes, one sliding into the other. The outer tube, with its row of holes along the axis and closed end, resembled the familiar penny whistle in appearance.

Odorant material was used to coat the inner surface of the outer tube. The odor evaluator then placed one end of this assembly into a single nostril and, with the outer tube set so that only one hole provided an air stream, inhaled to attempt perception of the odor. Sliding the outer tube away from him successively exposed more holes for inspired air and thus a greater surface for odor entrainment from the exposed inner wall of the outer tube.

Odors were rated 1, 2, 3 . . . according to the graduated distances along the outer tube that had to be exposed to produce a positive odor response (i.e., odor just perceptible).

A modified version of the Zwaardemaker olfactometer, designed in 1917 by Van Dam,[1] substituted for the inner tube a rod prepared from the actual odorant material dissolved in paraffin.

Subsequent development of instruments for determination of odor threshold followed roughly this same approach of varying the amount of dilution air while keeping the mass of odorant constant. The "osmoscope" of Fair and Wells,[2] devised in 1934, was used rather effectively to evaluate sewer air by successive dilutions using equal volumes of fresh air to achieve an odor threshold value. This approach was probably the first to employ units of 1:1 dilution volumes on the premise that such gross changes in the odor intensity are more easily recognized by the subject than the slight odor level changes inherent in earlier devices. Here, the number of equidilutions required to eliminate the odor were recorded and expressed directly as the pH.

ODORS AS AIR CONTAMINANTS

An odor may be defined as any stimulus that is perceived by the sense of smell. It becomes a problem as a community nuisance when it produces physical discomfort to a reasonable number of residents and interferes with their reasonable comfort and sense of well-being.

Evaluation of such a pollutant proceeds from the identification of:

1. Odor quality
2. Odor intensity

Odor quality is most often designated by using a scale developed by Crocker and Henderson,[3] which divides quality characteristics into four categories—fragrant, acid (acrid, biting), burnt, and caprylic (ripe cheese). These four qualities are rated individually on a scale of 0 to 8, and the overall quality is described by a four-digit number. For example, an odor such as butyric acid (rancid butter) might be described as 2106. This quality expression represents:

2—Slightly fragrant
1—Slightly acidic
0—No burnt odor
6—Highly caprylic

Odor intensity is best described as a physiological response to the concentration of a particular odorant. This phenomenon has been expressed in terms of a mathematical relationship known as the Weber–Fechner law. Weber first noticed that, in the human response to odor, a certain minimal amount of stimulus is required to produce a sensation. Fechner postulated further that equal increments of stimulus produce equally more intense sensation. Expressed mathematically, $\Delta R = K(\Delta S/S)$, or the change in response R is directly proportional to the ratio of the change in stimulus ΔS to the stimulus.

Integrating this expression gives $R = K \log S$. Translated into words, this means that, in a given situation of sense stimulation such as odor preception, the magnitude of response is directly proportional to the log of the stimulus.

In using this relationship, it must be emphasized that, like other logarithmic relationships (e.g., Beer–Lambert law), this law holds best over middle ranges of sense intensity and breaks down at extremely high or low concentrations of odor. The suggestion here is that the organism cannot satisfactorily come to equilibrium with either gross or threshold concentrations of detectable odor. Thus a reasonably constant slope, K, is not achieved at these concentrations.

Sources of Odors in a Community

As a rule, community-related odor nuisance problems are most often associated with organic compounds. Table 1 shows a number of common sources of community-related odors, together with the chemical composition of the odor and the general olfactory characteristics.

THEORY OF ODOR PRECEPTION

When stimulated by odorous molecules, olfactory nerve endings signal the more complex brain centers, where these stimuli are interpreted in terms of the kind and intensity of the odor.

This action-reaction-recall sequence requires that the odor sensed must have certain basic physical, and perhaps molecular, properties. Among its physical properties, the odorant must be rather low boiling to vaporize into the sense organs. Therefore, no odor should be expected from pure metals, most minerals, and even heavy paraffins. In addition, the odorant should be rather water soluble, or at least miscible in the mucous lining of the nose.

In addition to these necessary physical properties required to permit the function of sensation, several researchers have suggested that it is the chemical conformation of the actual molecules that produces the variety of odors observed by man. For example, the odor intensity of perfume appears to be a function of the branching of a straight-chain compound.[5] Floral odors seem to arise from the smaller more spherical molecules, and distortion of the sphere with bulky groups tends to produce a sweet but muskier odor. Like ethyl ether, with its balanced linear band, most ethereal smelling compounds (e.g., methylene chloride, dioxane, or diethyl ether) appear to be somewhat symmetrical.

In 1962, Amoore[6] published a theory that accounted for seven primary types

Table 1 Sources and Characteristics of Odorous Compounds

Source	Chemical Compounds	Olfactory Characteristics
Dry cleaning	Ethylene dichloride	Ethereal
	Methylene chloride	Ethereal
	Mixed hydrocarbons: standard solvent	Kerosene
Tallow rendering	Alkanes, C-5 to C-8	Kerosene
	Alcohols, C-2 to C-5	Fragrant alcoholic
	Aldehydes, C-2 to C-9	Pungent, gardenlike
	Ketones, C-3, C-4	Fragrant, airplane cement
	Acrylonitrile	Medicinal
	2-Methyl tetrahydrofuran	Fragrant
	Toluene, benzene	Fragrant, aromatic
	Dipropyl sulfide	Decaying cabbage
Asphalt	Terpenes	
	Hydrogen sulfide	
Starch frying	Oleic acid	Rancid fat
Food processing	Butyric acid	Rancid butter
Acid storage plants	Nitric acid	Sweet pungent
	Hydrochloric acid	Acrid biting
	Sulfuric acid	Acrid biting
	Acute acid	Vinegar
	Phosphoric acid	Acrid biting
Oil refineries	Hydrogen sulfide	Odor of rotten eggs, nauseating
	Thiophene	Putrid, nauseating
Combustion of fuel	Acrolein	Pungent, lacrimator
	Phenols	Medicinal
	Formaldehyde	Medicinal, irritating
Paint manufacturer	Methyethyl ketone	Fragrant
	Ethyl acrylate	Medicinal, pungent
	Methyl acrylate	Medicinal, nauseating

of odors in terms of their stereochemical classification. These classes were:

Ethereal   Pungent
Camphorous Minty
Musklike   Putrid
Floral

He suggested that a relationship existed between the geometry of molecules producing these classes of odors so that pungent-smelling molecules all produced a similar olfactory sensation because of their charge. This could be interpreted as a consequence of a charge separation on the part of the sense organ induced by the polarity of the odorant molecule. We can see that this theory holds true if we consider the structure of

$$^+H::::Cl^-$$

with its unsymmetrical charge distribution, which is very similar to SO_2 and NO_2,

$$^{O+}S=O^{O-} \quad ^{O+}N=O^{O-}$$
$$\parallel \qquad\quad \parallel$$
$$O \qquad\quad\; O$$

and also rather acrid-smelling molecules that "bite" the nose when inhaled. Compared to the symmetrical CO_2 (O = C = O), which is odorless, one can envision how these angular molecules without charge symmetry may well actuate olfactory sites that are not keyed to entry by other molecules.

Some examples of the "molecular conformity" have been suggested by Amoore, as shown in Figure 1. Curiously, optical isomers (molecules that are quite similar chemically) have been found to possess quite different odors, which suggests again the "keyed" rather than chemical quality of odor stimuli.

The shapes of the above molecules were determined using x-ray diffraction, infrared, and electron probe methods. Where the molecule appears to have a shape as shown, it may be presumed that the arrangement of nerve endings at the sense organs is a template structure conforming to the size, shape, and perhaps the charge distribution of the odor molecule. In this respect, as mentioned previously in regard to HC1, the class of acrid and putrid odors appears to be more dependent on polarity and charge distribution than on bulk molecular conformity, since molecules that stimulate this type of response are very small and have an unbalanced charge distribution. Thus NH_3, H_2S, PCl_5, and NO_2 are observed to fall into this category of acrid, putrid odors, which correspond to a compact, but charge-asymmetric molecule.[7]

Molecular size or diameter also appears to be a criterion for relating some of the more bulky molecules such as the examples in Figure 2, where each of the molecules shown has a molecular diameter of approximately 7 Å.[8]

This should suggest to the reader who may be unfamiliar with the odors of various chemicals that it is possible to classify an odor as one of the seven primary odors:

1. Musky
2. Camphorlike
3. Floral
4. Peppermint

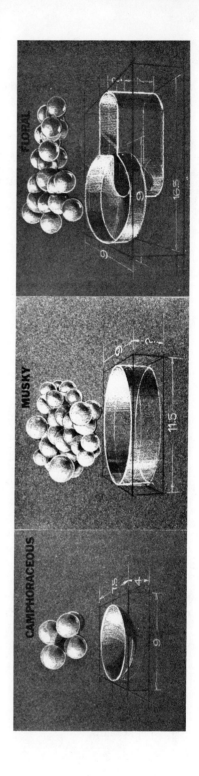

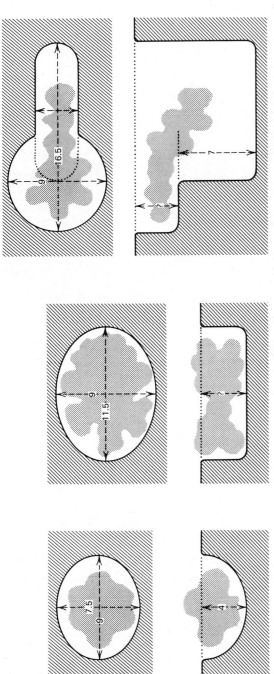

Figure 1. Receptor sites for three of the primary odors and example arrangements of odor molecules. From "The Stereochemical Theory of Odor" by J. E. Amoore. Copyright © 1964 by Scientific American Inc. All rights reserved.

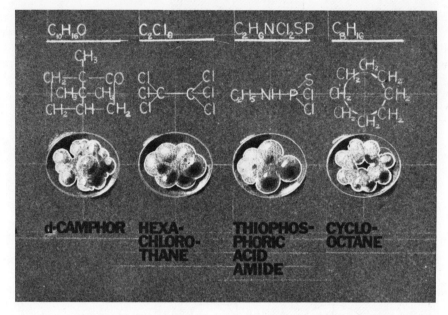

Figure 2. Compounds that are structurally unrelated but have a spherical 7-Å molecular diameter and a camphorlike odor. From "The Stereochemical Theory of Odor" by J. E. Amoore. Copyright © 1964 by Scientific American Inc. All rights reserved.

5. Ethereal
6. Pungent
7. Putrid

by means of some knowledge of the chemical structural formula together with an idea of its polarity.

ODOR THRESHOLDS

Because of the subjective character of odors, the concept of odor threshold has become a function of consensus rather than absolute measurement. The literature in the area is spotted with references based on various olfactory techniques, each of which claim to remove as many as possible of the subjective factors in an attempt to yield reproducible results, when applied to any group or panel of odor evaluators.

Probably the earliest attempt to compile a comprehensive table of odor threshold limits was undertaken by Katz and Talbert in 1930.[9] The olfactory technique in this case employed a thermostated U-tube containing solid or liquid odors. Clean, dry air was passed through this tube before entering a plenum that served as a mixing chamber for dilution air, which was also dried and charcoal filtered. The mixture of dilution air and odor was then passed into a mask or nose cone through which the human evaluator inhaled and evaluated the odor intensity. This technique utilizes the advantage of vapor pressure data, known for the pure odor as a chemical species. to calculate parts per million flow in much the same way as a permeation tube (Chapter 6) is arranged with a flowmeter to yield standard concentrations of contaminant gases.

For example, CCl_4 at a partial vapor pressure of 100 mm at 230°C through evaporation produces a weight loss of 5 mg of CCl_4 in 105 min if the clean airflow is 1 cfm. From this, the CCl_4 concentration is

$$\frac{\text{Total weight of contaminant}}{\text{total volume of air}} = \frac{5{,}000\ \mu g}{10.5\ ft^3} \times \frac{35.31\ ft^3}{m^3} = 16{,}767\ \mu g/m^3$$

Converting this to volume (parts per million) gives

$$\text{ppm}_v = \frac{\mu g}{m^3} \times \frac{0.0245}{MW} = 2.83\ \text{ppm}$$

This metered level of 2.83 ppm can then be related to the vapor pressure versus temperature curve for carbon tetrachloride so that odor concentration can be adjusted by either airflow or temperature control.

Odor thresholds produced by this method using the Katz-Allison odormeter have been recognized for many years and formed the basis for thresholds published in the 1951 air pollution manual of the Manufacturing Chemists' Association.[10]

Subsequently, as a result of disputes over rather widely varying odor thresholds reported elsewhere in the literature, the Manufacturing Chemists' Association recalled this manual and, in 1967, submitted a research proposal to Arthur D. Little Inc. to determine odor thresholds for 53 recognized odorants using a technique that would have the authority of a reference method.

In the resulting procedure, the individual subject exposure approach was replaced by the concept of an "odor room" into which metered concentrations of odorant compounds would be ventilated. Trained odor analysts, each having experience in recognizing the character of various odors, were selected to produce a more rigorous threshold value that, of necessity, required each panel member to describe the quality of the odor. The resulting threshold was then defined:

. . . as the first concentration at which all the panel members have been able to recognize the odor sensation that is characteristic of the chemical and have been consistent in their response at all higher concentrations even when these are presented in random fashion to minimize sample order bias.[11]

The odor thresholds in Table 2 are the products of this study. In the belief of

Table 2 Odor Threshold and Characteristic of Common Odorants in Air Expressed as Parts Per Million by Volume[11]

Compound	Olfactory Characteristics	Odor Threshold (ppm by Volume)[11]
Acetaldehyde	Green, gardenlike	0.21
Acrolein	Burnt, pungent	0.21
Amine, dimethyl	Garlic, onion, pungent	0.047
Amine, monomethyl	Fishy, pungent	0.021
Amine, trimethyl	Fishy, pungent	0.00021
Butyric acid	Sour, rancid butter	0.001
Diphenyl sulfide	Burnt, rubbery	0.0047
Ethyl acrylate	Hot plastic, earthy	0.00047
Ethyl mercaptan	Earthy, sulfidy	0.001
H_2S gas (dry)	Rotten eggs	0.00047
Methyl mercaptan	Sulfidy, pungent	0.0021
Methyl methacrylate	Pungent, sulfidy	0.21
Paracresol	Tarlike, pungent	0.001
Phenol	Medicinal	0.047
Pyridine	Burnt, pungent, diamine	0.021
Toluene diisocyanate	Medicated bandage, pungent	2.14
Trichloroethylene	Solventlike	21.4

the author they are the most reliable presently available in terms of the limited number of human variables permitted in this study.

As one might expect, as a group, sulfur compounds shared the olfactory distinction of the lowest threshold levels. In fact, only H_2S and SO_2 showed levels above the range of parts per billion.

The lowest measured threshold was found to be that of trimethylamine at an astoundingly low 0.21 ppb. Many nitrogen-containing compounds, such as common proteins and amides, showed much higher thresholds up to 100 ppm for dimethylformamide.

Oxygenated compounds were spread over a wide spectrum of odor thresholds, as were unsaturated and benzene-substituted compounds. However, in the case of the aromatics, substitution of an otherwise pure ring structure was observed to lower the odor threshold by factors of 10^{-2} and 10^{-3} per substituent, depending on the type of group.[11] In this respect, electrophilic groups appear to lower the thresholds to a greater extent than a nucleophilic substituent, although the low odor threshold of nitrobenzene may be due to its unique and easily recognizable sweetish odor when exact identification of the odor forms an important part of the panelist's odor response.

COMMON INDUSTRIAL ODORS

H_2S

No routine measurements are made for H_2S by the National Air Pollution Control Office of the Environmental Protection Agency.

Limits on this offensive and, in high concentrations, deadly gas are established and measurements are made by local communities where a local source is known.

Hydrogen sulfide occurs naturally from such sources as sulfur springs, stagnant water, and natural gas and oil.

Sources that yield H_2S as a product of man's actions include coke ovens, Kraft paper mills, iron smelters, the chemical and dye industry, and petroleum refineries.

Many industries produce captive amounts of this gas, which are stored for reclaiming or are "flared" to the atmosphere by burning to emit the less objectionable SO_2. However, it is not always the known process leak that becomes a source of community odor problem. Instead, the source may be process wastes that are drained off into storm sewers. This fugitive source may produce odor miles from the actual source, thereby producing a rather difficult problem for assignment of responsibility, which is a necessary prerequisite for abatement.

Several approaches to H_2S sampling are presented under determinations of gaseous pollutants in Chapter 3. However, of the means available to detect this odor, the human nose is by far the most sensitive and, in the last resort, the final judge of objectionability.

Odor thresholds for H_2S have been published since 1939, as a result of the work of Dalla Valle[12] and McCord.[13] These, together with some more recently determined levels, are presented in Table 3.

Beyond the pure application of numbers to the concept of odor threshold is the necessary correlation factor of "odor gradient," which attempts to assign qualitative differences to various concentration intervals as an aid in distinguishing degree of response once the real odor threshold itself has been exceeded.

The gradient of odor response for H_2S is identified by the *Preliminary Air Pollution Survey of Hydrogen Sulfide* (U.S. Department of HEW) 1969.[17] Here, it was found that the odor is "distinct" at 0.350 ppm, becomes "offensive and moderately intense" and is "strong, but not intolerable" at 21 to 35 ppm.[17,18]

At 226 ppm, the odor is not as pungent as at lower levels because of the chemical or physiological reaction that causes paralysis of the olfactory nerves.[19] Threshold concentrations are based upon fresh low level inhalations. However, repeated inhalation causes loss of perception or olfactory fatigue so

Table 3 Odor Detection Thresholds for H_2S

H_2S Threshold (ppm)	Date	Authors
0.0065–0.0324	1965	Adams and Young[14]
0.002	1966	Larson
0.005	1968	Wilby[15]
0.008–0.02	1959	Fyn-Djui[16]
0.7	1939	Dalla Valle[12]
0.7	1949	McCord and Witheridge[13]

that, at concentrations above 7840 ppm, there is "... practically no sensation of odor ... death can occur rapidly."[20]

On the other hand, it is also well known that, at concentrations below 0.1 ppm, probably the only serious objection to H_2S results from its recognition as an odor nuisance.[21] Consequently, since man's peace and sense of well-being are closely tied to his olfactory perturbation, abatement of this odor so that no threshold level exists in the community is a worthwhile aim. Recognizing this fact, a realistic air standard for H_2S should lie between 0.001 and 0.009 ppm, and such a level is measurable by the Methylene Blue Method described in Chapter 3.

Rendering Odors

Whereas H_2S is considered a toxic gas as well as one whose odor is disagreeable, gases and vapors emitted from the rendering of inedible animal products are limited to the area of nuisances and do not produce generally recognized toxic effects. However, exposure to extremely high concentrations of these rendering effluents can produce nausea of the same order as experienced upon exposure to rancid food odors in the home. Therefore, the mere disagreeability of these odors cannot be considered sufficient reason to overlook this rather important source of air pollutants.

Individual sources of odor within the rendering process center around:

1. The raw dumping area where carcasses are transferred from the tarpaulin-covered trucks to a hopper or bag from which they will enter the grinding and cooking operations. Relatively low odor concentrations are usually evolved from this area.

2. The cooking process in which animal parts that have been "hashed" are heated in a steam-jacketed vessel or are contacted directly with live steam in an

enclosed container. Temperatures of 225 to 300°F are required to decompose calcium linkages in the connective tissue to free tallow, which is the salable product. Since this portion of the rendering process usually produces the most disagreeable odors, it is essential that gaseous effluents be scrubbed with a reactive solution, such as potassium permanganate (3 to 6%), or incinerated before entering the atmosphere.

3. The tallow storage, which usually requires silos, produces moderate odor levels, especially during the filling cycle in which odorous air is displaced from an empty or partially empty silo tank.

Other related odor source areas include the blood concentration, the press area (where tallow is squeezed from the bone "cracklings"), and the grease-drying operation.

In an effort to determine the relative contribution of these point sources to the general problem of rendering-plant odor, the Fats and Proteins Research Foundation sponsored a research project in which data were produced using the relative flame ionization response of a hydrocarbon detector to C_2–C_{29} compounds. The resulting contribution of various sources to odor are shown in Table 4.

Using gas chromatography together with mass spectrometry, actual chemical components of the rendering effluent were identified. A partial list of these compounds is presented in Table 5. The major objectionable odor components are aldehydes, ketones, and carboxylic acids. It is essential that any proposed control program for abating these odors by chemical means be directed along the lines of collecting these compounds by their specific chemical reactivities (i.e., aldehydes by bisulfite, ketones and acids by potassium permanganate solution).

Specific determinations for aldehydes, ketones, and carboxylic acids are

Table 4 Relative Contribution of Plant Sources to Rendering Odor[22]

Plant Area	Odor Component Concentration (μg/liter)
Press ventilator	27.4–50.4
Press area	4.5– 8.5
Raw-stock area	1.9– 3.8
Cooker dump area	3.1– 5.8
Cooker-condenser	10.0–11.2
Grease-drying operation	7.7– 9.6

Table 5 Some Components of Rendering Odor[23]

Acetaldehyde	Isovaleric acid
Acetic acid	2-Methylpropanol
Acrylonitrile	2-Methyltetrahydrofuran
Benzene	3-Methylbutanol
1-Butanol	n-Pentane
2-Butanone	n-Propanol
Butyric acid	Propanal
Ethanol	

found in Chapter 2, while principles for separation methods required to concentrate samples in these compounds are included in Chapter 3.

Pulp and Paper Process Odors

Since the most prevalent odor related to pulp processing is H_2S, the reader should refer to the previous section on H_2S for information concerning this toxic and malodorous compound. Other compounds produced by the action of sodium sulfide and sodium hydroxide on wood chips (Kraft process) are mercaptans, organic sulfides, and organic disulfides.

Specific compounds emitted at particular process points in the Kraft pulp process have been determined by Sableski. The more pronounced of the odor points and the measured concentrations of the known odorous compounds at these points are shown in Table 6.

Table 6 Range of Sulfur Gas Concentrations Encountered in Kraft Mill Sampling[24]

	Concentration (ppm)			
Source	Sulfur Trioxide	Hydrogen Sulfide	Methyl Mercaptan	Dimethyl Sulfide
Digester vent	—	16–18,800	0–4,370	3,850–65,000
Evaporator	—	907–32,600	455–36,700	0–27,600
Recovery furnace	4–798	14–1,140	0–489	0–260
Tall-oil cooking	2–822	5,400–101,000	0–4,660	0–0

When paper and pulp processing is being surveyed for odor levels, using the above information on specific compound concentration, it is advisable to determine the ambient concentrations of these pollutants so that comparisons with odor thresholds can be made. Methods for the sulfur-related compounds above are presented in Chapter 3.

Petrochemical Odors

Again, H_2S is probably the most prevalent odor, although other "crude oil" odors (e.g., phenols, naphthenic acids, organic sulfides, organic amines, mercaptans, and aliphatic or aromatic aldehydes) are also detectable chemically if not always olfactorily.

A list of these compounds, as presented by the Petroleum Committee for the Air Pollution Control Association, is reproduced in Table 7, along with a partial description of their sources.

Paint Manufacturing Odors

Probably the most offensive odor produced in the preparation of paints is the acrylic odor resulting from polymerization of ethyl and methyl acrylates during the synthesis of acrylic latex. Temperature control is essential in this process; an increase of 2 to 5°F during the first stages of polymerization can result in the evolution of upward of 10 to 20 pounds of ethyl or methyl acrylate, where

Table 7 Potential Sources of Odorous Emissions from Oil Refineries[25]

Emission	Sources
Oxides of sulfur	Combustion of sulfur-containing fuels, flares of H_2S containing gases, catalyst regenerators.
Hydrocarbons	Storage tanks, oil recovery units, catalyst regenerators.
Oxides of nitrogen	High-temperature combustion processes.
Mercaptans	Cracking processes, asphaltic cooking, cracking, and separation.
H_2S	Hydrosulfurization and hydrocracking units.
Phenolic compounds and naphthenic acids	Effluent from caustic scrubbing.
Organic sulfides and nitrogen bases	Effluent from acid scrubbing solutions.
Aldehydes	Partial or low-temperature combustion of fuel, cracking of aromatics, air blowing of asphalts.

the odor threshold of these compounds is on the order of 0.00047 ppm by volume. Quantitative analysis for these compounds can best be achieved by gas chromatography of an air sample concentrated by cryoscopic or impinger techniques. Good retention is achieved on a silicone column, and pure standards dissolved in carbon tetrachloride can be employed to determine retention time.

Community Odor Complaints

Preliminary to a discussion of dilution methods of odor evaluation, it would be appropriate to note conditions under which odor is most frequently the subject, since these aspects will determine the accuracy of the complaint prior to sampling for the odor.

In respect to the effects of meterological conditions on the retention and subsequent buildup of odor levels from an otherwise "dispersing" source, such as a smoke stack, a study was made by Huey in 1959 to determine the most influential atmospheric effects in odor pollution problems. As might be expected, fewer complaints of odors were noted in the winter months when, at least in the northern states, temperatures below 40°F resulted in condensation of gaseous odorant at or near the source with only minimal diffusion of vaporous pollutant to the community at large. On the other hand, low to moderate winds tended to favor a slight increase in the number of complaints. Increasing humidity produced a decrease in odor complaints, but an increase in the number of exposure hours per complaint, as shown by Table 8. Here, it would be assumed that water vapor exhibits a dilution effect on gaseous odors. Since higher-humidity air is less dense by reason of the low molecular weight of H_2O,

Table 8 Effect of Relative Humidity on Odor Nuisance Occurrences[26]

Relative Humidity	Number of Complaints During 1958–1959	Number of Exposure Hours at RH	Ratio: Number of Exposure Hours Per Number of Complaints
0– 30	0	435	∞
30– 49	27	2974	110
50– 69	47	5186	110
70– 79	24	3184	132
80– 89	18	3007	167
90–100	18	2698	150

this air is more buoyant and tends to disperse odors more rapidly. It could also be postulated from the steric odor theories of Amoore and of Moncrieff that a positional blockage or preferential absorption takes place in the olfactory sense organ by reason of which nonodorous water molecules effectively mask the effects of odorous molecules.

METHODS OF ODOR MEASUREMENT BY DILUTION

Dilution methods for odor determination require the quantitative dilution of odorous air with odorfree air by successive batch dilutions until panel subjects exposed to these dilutions can no longer smell the odor.

Quite a number of methods are available for bringing the nose of the panelist into contact with the measured odor–air mixture. The mixture may be made in a container, such as a Mylar bag, and the panelists allowed to inhale odor from the bag as one of several batch bag dilutions. The odor–air mixture may also be conveyed into a nose mask by injection; however, this strategy is discouraged because of the possibility of a nonhomogeneous mixture of the odor with the dilution air existing in the cavity beneath the panelist's nose. The most generally reliable method for measuring the odor strength from source samples is probably the ASTM syringe method (ASTM D 1359-57).

Syringe Method

In this method (ASTM D 1359-57), a glass sample pipette of at least 250-ml volume is filled with odorous air at its source by means of a one-way squeeze bulb attached to the stopcock opposite the source (Figure 3). At least 60 air changes must be drawn through the pipette to ensure representative sampling. Alternatively, a portable diaphragm pump attached in the same way as the rubber bulb may be used to fill the pipette.

At some time just before or after the odor sample is withdrawn, the velocity of air in the stack or duct sampled should be measured and recorded using a Pitot tube, attached to an inclined manometer or a Magnehelic flow gauge.

One or more samples of odor so obtained are then returned to the laboratory where they are kept at room temperature until an odor panel can be assembled. It is essential that these samples be evaluated within one day's time to ensure reliable results. Some odors, especially carbohydrate fumes, are known to adsorb onto a glass surface with time, and decreasing odor strength has been noted with standing time in excess of 12 to 24 hr.

A number (10 to 12) of dilution syringes of 100-ml capacity can then be prepared using a small 10 to 50 ml gas-tight transfer syringe (Figure 4) to

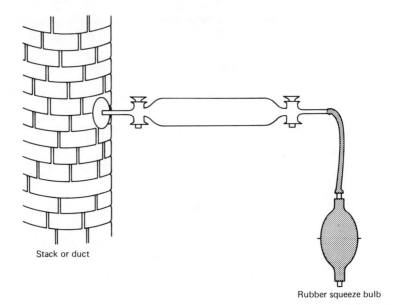

Figure 3. Diagram of odor sampling at the odor source.

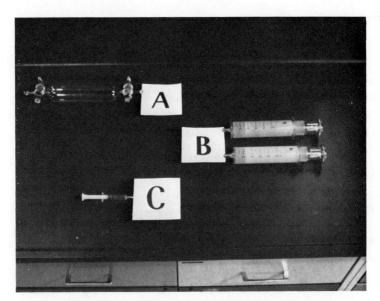

Figure 4. Equipment for odor level by the dilution method.

withdraw measured volumes of odor, which are then used to partially fill the 100-ml dilution syringes.

Procedure

Using a transfer syringe (c), withdraw a 10-ml quantity of odor from the sample syringe (a). Keeping the plunger tightly secured to the 10-ml mark, insert the needle tip of the transfer syringe into a 100-ml dilution syringe (b), which contains some odorfree air. Inject the 10-ml aliquot, withdraw the transfer syringe, and fill the dilution syringe to the mark with odorfree air. Allow 15 to 30 sec to pass before using the dilution syringe, which contains a 9:1 dilution, for evaluation by members of the panel. Similar dilutions of 1:1, 2:1, 5:1, 20:1, 50:1, 100:1, 500:1, 1000:1, and 2000:1, are common, and each dilution syringe may in turn be filled in the above manner and exposed to the panel.

When a sufficient number of dilution syringes is not available for the desired dilution levels, an alternate procedure can be used. Measured volumes of a given dilution mixture (e.g., 10 ml) are injected into the nostril of each panelist, as shown in Figure 5. After exposure of 50 ml to a total of five panelists, the remaining 50 ml of given dilution mixture are used by refilling the syringe to the mark with odorfree air and noting this subsequent dilution as a 2:1 ratio of the initial dilution.

Panel members are exposed to successive dilutions, preferably in random order. At least one out of every four to five dilutions should be a blank, a decoy, or a "scramble" (prepared from an unrelated dilution) to prevent panelists from anticipating a lower or higher dilution. Thus the panelist must be alert only to his own positive or negative response to the proper dilution. Responses to these nontest related dilutions are noted and these results are deleted from the panelists' charts (Figure 6) on which they keep track of their responses by marking Y (yes) if the member perceives the given odor dilution (marked A_1, B_4, C_6, etc.) or N (no) if he fails to perceive it.[27]

The chemist or technician who is conducting the panel test continues to expose panel members to these successive, but random, dilutions until at least three-fourths of the panel can no longer detect the odor. A panelist is considered "out" when he fails to perceive two samples of successively lower dilutions which differ in concentration by a factor of 2:1. However, neither he nor the other panel members are so informed, and the panel is continued until the final panelist has failed to perceive three successively lower dilutions.

The Panel

In reviewing the literature, the reader may find many different approaches to evaluating subjects for a prospective panel and to studies related to the panel itself.[28-30] Here, a study of the degree of variance in panel response conducted by

Figure 5. Chemist, not a part of the panel, injects a measured volume of diluted odor into the nostril of panel member.

Prince and Ince[31] showed marked differences in the abilities of panel members to respond to various odors. In addition, such variables as diet, environment, and daily outlook (mood) seemed to play a part in a panelist's response in degree or lack of ability to perceive a given odor level. In general, for four to five trained panelists (chemists or persons familiar with chemical odors), results were reproducible to approximately 20%.

The generally accepted number of panelists varies from 5 experienced persons to 11 inexperienced persons, with roughly the same reproducibility achieved at either extreme where as many as 35 panelists were reported in one program of evaluation.[32] Most references, however, proceed with odor evaluations using four to five panelists who have demonstrated the ability to distinguish odors according to a simple screening test in which two different odors must be differentiated at successively lower concentrations.[33]

Calculations

Raw data are assembled by the panel moderator from data sheets that have been distributed to each panelist. An example of such a sheet is shown in Figure 6. As a rule, no more than three samples are presented to the panel at one sitting.

The odor threshold for the panel represents the point at which ". . . the difference between the greatest dilution at which odor is consistently perceived and the next greatest dilution measured is less than 50 percent of the greatest dilution at which it is consistently perceived."[34]

This value is obtained graphically by plotting the log of the lowest odor dilution ratio at which each panelist reported positive olfactory response versus the percentage of the panel reporting positive response.

A typical plot of this kind is shown in Figure 7, which shows results when odor concentrations of the storage tank vent and the processing room incinerators in a typical tallow-rendering operation were measured using the syringe method.

A good means of odor concentration comparison can be obtained from operating parameters of rendering processes. Fume incineration effectiveness

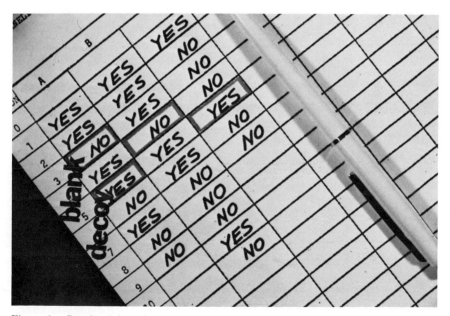

Figure 6. Panelists' data sheet.

depends on residence time, flame temperature, and volatility of the incinerate. For example, an increase in flame temperature was observed to produce a remarkable decrease in odor concentration, as shown in Figure 7.

Data representing this odor evaluation are found in Table 9, where it is seen that an increase of only 30 to 50% in Fahrenheit temperature produced an odor reduction on the average of 156:1.

Here, the emission rate in odor units per minute is calculated by multiplying the static value of odor units (ou) per cubic foot by the flow in cubic foot per minute:

$$\frac{1450 \text{ ou}}{\text{ft}^3} \times \frac{220 \text{ ft}^3}{\text{min.}} = 319,000 \frac{\text{ou}}{\text{min.}}$$

Actual static odor unit concentration per cubic foot is determined graphically as shown Figure 7, where tracing the point of 50% panel response (arrow) to the ordinate yields the mean population dilution threshold in odor units, where the odor unit is the number of dilutions required to dilute 1 ft³ of odorous air to 1ft³ of air at the odor threshold (ASTM D 1391-57).

It has been found through experience that approximately 40 to 150 ou/ft³ constitutes a safe static level of odorant—one that produces no community com-

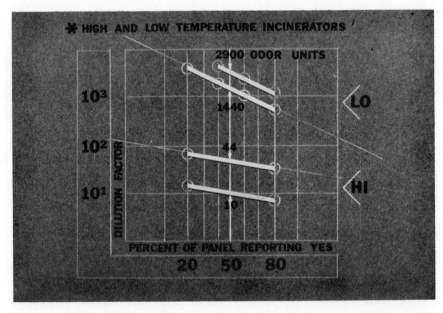

Figure 7. Plot of olfactory response versus log of the dilution factor for rendering incinerators at low (a) and high (b) flame temperatures. (From Ref. 27.)

Table 9 Odor Concentration Comparison of the Effect of Temperature Increase in the Effectiveness of Rendering Odor Incineration

Incinerator Sampling Point	Odor Units Per Cubic Foot at 50% Panel Reporting Positive Olfactory Response	Sampling Temperature (°F)	Flow (cfm)	Odor Emission Rate, (ou/min)
Storage tank vent[a]	1,440	850	220	319,000
Processing room[a]	2,900	650	2,000	4,200,000
Storage tank vent[b]	44	1,100	220	9,680
Processing room[b]	10	950	2,000	20,000

[a] Low-temperature incineration.
[b] High-temperature incineration.

plaint problem. At the same time, a high emission rate of such an odor (e.g., flow of 10,000 cfm) would result in an output to the community of as many as 1,500,000 ou/min. Here, experience has shown that a high emission rate of odor (i.e., in excess of 1,000,000 ou/min) can produce a nuisance level of odor in the community even though the static odor content of the emission is acceptable.

Notes on Interferences in the Dilution Method

If care is not exercised in the selection of an odorfree room, extraneous fugitive odors may influence the outcome of the test. This area should also be clean and quiet.

If the panel moderator has come in contact with the odor through field sampling, he should make every effort to reduce the subtle influence of lingering odors on his hands or clothing.

The panel odor observers themselves should avoid eating, drinking liquids, smoking, and chewing gum for at least 30 min prior to the time of test. In addition, any panel member who is aware of a physical limitation in himself, such as a "cold" or other respiratory problems, should announce this to the panel coordinator prior to the test and request that he be replaced by another qualified panelist.

Syringes as well as sample pipettes should be scrubbed thoroughly with unperfumed detergent and rinsed with tap water before being used in odor evaluation.

When all necessary precautions have been exercised, the actual precision and accuracy of this method depends on the number of observers and their acuteness in perceiving the odor. Based on the reproducibility of $\pm 50\%$ expected of any single panelist, reproducibility approaching 10 to 20% should be achieved by a panel of five or more observers.

RELATING ODOR CONCENTRATION TO CHEMICAL-PHYSICAL MEASUREMENTS

The problems of matching odor unit values to chemical or physical measurements designed to detect some one or a class of chemical compounds are numerous, and the approach in general is difficult to resolve.

The main problem in attaching chemical measurements to quantitative measurement of odors is probably the overcoming of chemical complexity. For example, the chromatographic separation of lemon oil reveals some 20 chemical components,[35] each of which would show its individual odor threshold. In addition, the mutual reactivity of odors, with various tendencies of the components to oxidize or reduce, adds a degree of "drift" to their standardization. Also, the very fact that odorous compounds often need be present in only trace quantities (i.e., parts per billion) to cause a community odor problem limits their analysis almost exclusively to mass spectrometry or gas chromatography.

It is this second instrumental option that we discuss briefly as a rather limited, but overall useful, approach to at least an evaluation of the effectiveness of odor control devices.

Some research in this area has been undertaken by the Rendering Institute, by Fuller, Stellenkamp, and Tisserand,[36] and by Trieff et al.[37] from the standpoint of actual odor abatement. The most positive practical results are found in investigation of efficiency measurements where the chromatograph serves to indicate the class or boiling fraction of compounds that evade a particular design of collection equipment.

Here, it is certainly not outside the scope of a class in air-industrial hygiene to collect a sample of industrial odorant using an inert collecting bag or gas pipette and to use standards such as formaldehyde or stearic acid to gas chromatograph the sample to determine the approximate boiling range of the uncollected effluent.

Very often, when light hydrocarbons appear in the effluent, refractory baffles can be installed and burners arranged to prevent the "tunneling" of these low-molecular-weight gases through the flame chamber of a fume incinerator.

When heavy hydrocarbons are in evidence, increase in flame temperature (e.g., 1300 to 1650°F) may well reduce the concentration of this higher autoignition point fraction.

Collection of Samples

A 10 to 12-in. length of Pyrex glass tube having an inside diameter of ⅜ in. is bent to a U-shape and packed with a hydrocarbon-active absorbent, such as Apiezon on Chromosorb W[36] or Carbowax on firebrick.[38] The tube is then attached downstream of a $CaCO_3$ or K_2CO_3 drying tube and immersed in liquid N_2, acetone-dry ice, or a salt ice freezing mixture while odorant air is passed through the column at 1 to 2 liters/min. Usual sample collection time is 1 to 2 hr.

The sample, which is allowed to remain in the freezing mixture until ready for chromatography, is then fitted with a fresh predrying tube and attached directly to the column of the chromatograph for separation.

The column separation can be achieved at room temperature to 100°C, with low temperature used at the outset when the more volatile components begin to evaporate into the air stream. Here, a programmed temperature oven with dual separation columns is especially useful in setting a temperature rise from 20 to 100°C to occur during ½ to 1 hr.

The volatility of most compounds separated from typical industrial odors (e.g., incinerator, asphalting, rendering, baking) is in the range from 0.5 times that of pentane in the case of lighter constituents to 8 to 10 times that of pentane for the heavy aliphatics and asphaltines.

Quite a number of substrate compositions can be used to pack chromatographic columns for this separation. Dinonyl or dibutyl phthalate, and Carbowax, silicon, or Apiezon-L on Chromosorb W yield good separations for a wide variety of functional groups as well as straight hydrocarbon mixtures.

After this operation has been checked for reproducibility (several sample traps can be collected simultaneously or in sequence), an output variable can be introduced into the industry's particular odor-collecting system. For example, a small scrubber consisting of an impinger filled with 1 to 5% potassium permanganate or 1 to 3% sodium hypochlorite can be inserted in the sampling line between the source and the predrying tube. Alternately, an auxiliary burner in a stainless steel housing can be inserted in the same location.

Gas chromatograms of the effluents in either case could serve to indicate the effect of such a design change on a pilot level.

Such a comparison is shown in Figure 7, where the effluent of hydrocarbon from a tallow rendering operation showed considerable odor reduction when the temperature output from an existing fume incinerator was increased.

The Kovats' Index System

If an indexing of individual components is desired, the Kovats system[39,40] is probably the simplest method for identifying the position of individual chromatographic peaks for more exhaustive characterization at a later time.

This system consists of a logarithmic transformation of the retention times of unknown compounds on a scale set for normal alkanes, after correcting for the retention time of methane. Here, the indices of the standard alkanes are assigned values of 100 times their number of carbon atoms (e.g., $K = 500$ for n-pentane and $K = 1200$ for n-duodecane.

Probably the most useful application of this formal numbering system arises out of the ability of the index to aid in determining the relative polarity of unknowns by plotting their "observed Kovats' indices" (which vary directly with the retention times) in polar versus nonpolar separation columns. Such a plot is shown in Figure 8.

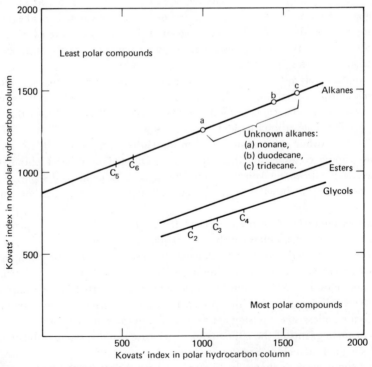

Figure 8. Polarity of unknown compounds can be determined and chemical class estimated by plotting the Kovats' indices in polar versus nonpolar gas chromatography columns. Unknown alkanes: (a) nonane, (b) duodecane, (c) tridecane.

In practice, the plot is established by determining the retention times of a series of known alkanes in each of two columns, such as Apiezon-L (representing a nonpolar separation medium) and Carbowax 20 M (representing a polar separation medium). Either the ordinate or the abscissa may be chosen to plot the observed Kovats' function of retention for these knowns in the more polar separation medium; however, the axis chosen then becomes the measure of polarity for unknown compounds that are chromatographed in the same columns and subsequently plotted on the same graph.

Here, a dual column analyzer would supply the differential data for such a plot, assuming at all times a value of 1 for the reference column that is plotted on the abscissa.

In the actual use of such a graph, the retention time, established in a polar medium on the abscissa, may be read up to the standard curve and over to the y-coordinate to establish a relative boiling point of the unknown with respect to alkanes plotted on this axis. Thus the positions of the polar derivatives of various homologs can be predicted or the polarity of a compound of known boiling point can be ascertained.

Although it is difficult to apply such a rigid numbering system uniformly to compounds that differ widely in chemical properties, because of the possible presence of several functional groups in the same compounds, a series of rules has been established to facilitate this comparison.

These rules may be summarized as follows:

1. In a normal homologous series of C_n where $n > 5$, the Kovats' index increases by approximately 100 units for each —CH_2— group present in the compound.

2. The indices of multifunctional or asymmetrically substituted compounds, such as ABC, can be calculated by using as reference the compounds ABA and CBC.

3. When two structural isomers exhibit a difference of one unit in the nonpolar phase, a difference of five units will be observed in their boiling points.

4. For any given compound, the difference between indices of this compound in a polar column versus a nonpolar column is an additive function of the number of functional groups.

In practice, some allowances for rule four must be made for hyperconjugation and bond interaction effects, which may influence the otherwise additive effects of functional group substituents.

THE ODOR ROOM METHOD

When it is possible to isolate an entire room for use in odor evaluation, a better experience of the odor may be gained by the panel remaining seated in the room while background air is diluted with odorant at a measured level and emitted into the room until equilibration is achieved (after 5 to 10 air changes). Thus the entire panel is exposed to exactly the same level of odor at the same time.

Unquestionably, more sophisticated instrumentation and planning must come into play in the use of this type of odor evaluation.

Where this technique has been applied, some researchers have reported tentative results of odor thresholds 2 to 3 orders of magnitude lower than those observed by the syringe method. Here, speculation would suggest that consistent, but unmeasured, dilution air is inadvertently inhaled by the panelist as he evaluates odorant from the measured volume of the syringe. This volume of background air would be roughly the same for all panelists and would tend to raise the level of the mean threshold odor response.

AMBIENT ODOR METHODS

When offending odors are observed in ambient air and no opportunity is afforded to obtain a sample of the raw odorant because of lack of access to the stack or other process contingency, the chemist may have recourse to a scentometer.[41] This instrument represents an attempt to quantify an observer's response to a level of odor observed in the air volume within a building or in the open air. The device uses a charcoal filter to remove odor and thus provide a source of background air.

The nose is placed over small vent nosepieces, and dilutions ranging from 1 to 25 using odorfree air are produced by means of various finger combinations over access holes in the charcoal bed.

In spite of some studies[42] showing the applicability of this device to a community's air studies, it has been the experience of the Wayne County Air Pollution Control Agency that the scentometer has been of questionable value in measuring rather low levels of odor that have been the source of citizens' complaints.

This apparent lack of applicability to low-level ambient odors does not preclude the successful application of the scentometer for evaluating in-plant odors that often exist at several times the outside ambient level.

REFERENCES

1. Van Dam, "Odor Detection," *Arch Neerlandaises Physiol.*, **1,** 660 (1917).
2. G. M. Fair and W. F. Wells, "Estimation of Odour in Water," *J. New Eng. Water Works Assoc.*, **47,** 249 (1933).
3. E. C. Crocker and L. F. Henderson, "Analysis and Classification of Odors," **22,** 325–326 (1927).
4. D. M. Doty, "Chemical Nature of Rendering Odors," *Directors Digest, Fats and Proteins Research Foundation Inc.*, **80** (1971).
5. J. Von Braun and E. Anton, "Effect of Chain Branching on Odor," *Ber.*, **62B,** 1489 (1929).
6. J. E. Amoore, "Stereochemical Theory," *Perfum. Essent. Oil Rec.*, **43,** 321 (1952).
7. J. E. Amoore, "The Stereochemical Theory of Odor," *Sci. Amer.*, **210,** 42–49 (1964).
8. *Ibid.*, p. 46.
9. S. H. Katz and E. J. Talbert, *U. S. Bureau of Mines Report*, **480,** 1–37 (1930).
10. "Air Pollution Abatement Manual, Psychological Effects," Manufacturing Chemists Assoc., Washington, D. C., 1951, Chap. 5.
11. G. Leonardos, D. Kendall, and N. Barnarch, "Odor Threshold Determinations of 53 Odorant Chemicals," *J. Air Poll. Control Assoc.*, **19,** 91–95 (1969).
12. J. M. Dalla Valle and H. C. Dudley, "Evaluation of Odor Nuisance in the Manufacture of Kraft Paper," *Public Health Reports*, **J4,** 35–43 (1939).
13. C. P. McCord and W. Witheridge, *Odors-Physiology and Control*, McGraw Hill, New York, 1949.
14. D. F. Adams and F. A. Young, "Kraft Odor Detection and Objectionability Thresholds," Washington State University Progress Report, U.S. Public Health Service Grant, 1965.
15. F. V. Wilby, "Variation in Recognition Odor Threshold of a Panel," Proceedings Annual Meeting, Air Pollution Control Association, St. Paul, Minn. 1968.
16. D. Fyn-Djui, "Basic Data for Determination of Limit of Allowable Concentration of Hydrogen Sulfide in Atmospheric Air," *Gigiena Sanit.*, **24,** 10 (1959).
17. S. Miner, "Preliminary Air Pollution Survey of Hydrogen Sulfide," Litton Systems, Environmental Systems Division, prepared under Contract Number PH 22-68-25, U.S. Department of Health, Education, and Welfare, Public Health Service, 1969.
18. F. A. Patty, ed., *Industrial Hygienst and Toxicology*, Vol. 1, Interscience Publishers, New York, 1958, p. 896.
19. H. Petri, "The Effects of Hydrogen Sulfide and Carbon Disulfide," *Staub*, **21,** 63 (1961).

20. "Permissible Emission Concentrations of Hydrogen Sulfide," Subcommittee of Effects of Hydrogen Sulfide, Committee on Effects of Dust and Gas of the Verein Deulscher Ingenieure Committee on Air Pollution, 21 A. C. Stern, ed., *Air Pollution,* Vol. II, Academic Press, New York, 1968, p. 78.
22. T. A. Burgwald, "Identification of Chemical Constituents in Rendering Industry Odor Emissions," Report IITRI No. C8172 prepared for Fats and Proteins Research Foundation, Inc., Des Plaines, Ill., 1971, p. 13.
23. *Ibid.*, p. 36.
24. J. J. Sableski, "Odor Control in Kraft Mills," A Summary of the State of the Art., U.S. Public Health Service, National Center of Air Pollution Control, Durham, N.C., May 10, 1967.
25. "Petroleum Refining Industry-Air Pollution Problems and Control Methods," Informative Report No. 1, *J. Air Poll. Control Assoc.,* **14,** 51 (1964).
26. N. A. Huey et al., "Objective Odor Pollution Control Investigations," *J. Air Poll. Control Assoc.,* **10,** 441 (1960).
27. P. O. Warner, J. O. Jackson, M. L. Bird, and L. Saad, "Recent Advances in Odor Monitoring: A Rapid Method for Source Measurement of Odor Levels Using a Modified ASTM Dilution Method," *Proceedings, State of the Art of Odor Control Technology Specialty Conference,* Air Pollution Control Assoc. Pittsburgh, March 1974.
28. A. Turk, "Selection and Training of Judges for Sensory Evaluation of Intensity and Character of Diesel Exhaust Odors," Monograph, Environmental Protection Agency, National Air Pollution Control Office, Cincinnati, Ohio, 1967.
29. Manual on Sensory Testing Methods, ASTM Special Tech. Publ. No. 434, May 1968.
30. Correlation of Subjective-Objective Methods in the Study of Odors and Taste, ASTM Special Tech. Publ. No. 440, June 1968.
31. R. G. H. Prince and J. H. Ince, "The Measurement of Intensity of Odor," *J. Appl. Chem.,* **8,** 314–321 (1958).
32. F. V. Wilby, "Variation in Recognition Odor Threshold of a Panel," *J. Air Poll. Control Assoc.,* **19,** 96–100 (1969).
33. D. M. Benforado, W. J. Rotella, and D. L. Houston, "Development of an Odor Panel for Evaluation of Odor Control Equipment," *J. Air Poll. Control Assoc.,* **19,** 101–105 (1969).
34. "Standard Method for Measurement of Odor in Atmospheres," ASTM Designation D1371-51 reapproved 1967.
35. E. Shiftan and M. Feinsilver, "Practical Research in the Art of Perfumery," in *Recent Advances in Odor: Theory Measurement and Control,* R. L. Kuchner, ed., *Ann. N.Y. Acad. Sci.,* **116** 692–704 (1964).
36. G. H. Fuller, R. Stellenkamp, and G. A. Tisserand, "The Gas Chromatograph with Human Sensor: Perfumer Model" *Ibid.,* pp. 711–724.
37. N. M. Trieff, T. C. Erdman, S. N. Field, and E. L. Kershenbaum, "Analysis of

Industrial Odor Emissions," presented to Nineteenth Pittsburgh Conference on Analytical Chemistry and Applied Spectroscopy, Cleveland, Ohio, March 1968.
38. T. A. Bellar, M. F. Brown, and J. E. Sigsby, Jr., "Correlation of Subjective and Objective Odor Response," *Ann. N.Y. Acad. Sci.*, **116**, 567–575 (1964).
39. E. Kovats, "Similarities Between Structure and Gas Chromatography of Organic Compounds," *Z. Anal. Chem.*, **181**, 351 (1961).
40. E. Kovats, "The Kovats' Retention Index System," *Anal. Chem.*, **36**, 31A (1964).
41. C. W. Gruber, G. A. Jutye, and N. A. Huey, "Odor Determination Techniques for Air Pollution Control," *J. Air Poll. Control Assoc.*, **10**, 327–330 (1960).
42. S. W. Hortsman, R. F. Wromble, and A. N. Heller, "Identification of Community Odor Problems by Use of an Observer Corps," *J. Air Poll. Control Assoc.*, **15**, 261–264 (1965).

INDEX

Abrasives (particulates), 13-14
Absorbers, 122-127, 134, 197, 216, 220-222
Absorbing liquids, 218-219
Absorption methods, for CO, 99-100
  for CO_2, 107-108
  for sample collection, 216-222
Acetaldehyde, 61
  odor, 298, 302
Acetates, birefrigence, 28
  sources, 62
  spot test for, 22
Acetic acid, odor, 302
Acetone, 62
Acid storage plants, as source of odors, 292
Acrolein, 61
  methods of analysis, 84-85, 258-259
  odor, 292, 298
Acrylates, odor, 292, 298
Acrylonitrile, odor, 292, 302
Adsorption methods for sampling, 222-223
Aerosol reactions of hydrocarbons, 55-57, 63
Aerosols, secondary, 63
  separation techniques, 24-26, 190, 192-193
  standards for, 53, 268
  X-ray diffraction analysis, 36-44
  see also specific types
Aerosol vapor condensation meter, 192-193
Air contaminants, *see* Contaminants
Airflow measuring devices, 206-216
Air quality criteria, 52-53, 55, 130, 131, 267-268, 300
Air sampling, *see* Sampling
Alcohols, analysis, 86
  as contaminants, 60, 62, 63
  odor, 292
  solvent extraction systems for, 64

Aldehydes, 60-61
  concentrations by cities, 212-213
  health effects, 55
  methods of analysis, 83-86, 258-259
  odor, 292, 303
  in oil refineries, 303
Aliphatic hydrocarbons, 56
  methods of analysis, 68, 69, 258-259
  solvent extraction systems for, 64-67
  sources, 62
Alkanes, Kovats index, 314-315
  odor, 292
Aluminum, spot test for, 22
Aluminum oxide, 13
Amico-Bowman spectrophotofluorometer, 80-81
Amines, determination, 69-70
  health effects, 55-57
  odor threshold, 298
  solvent extraction systems for, 64-67
  sources, 63
Ammonium, concentrations by cities, 262-263
  sampling for, 254-255
Ammonium chloride, refractive index, 30
Anemia, 56
Anthracene, fluorescent spectra, 74
  sources, 62
Antimony, determination, 156
  sources and effects, 150
Aqueous solubilities of gases, 218
Aromatic colors, 71, 85
Aromatic hydrocarbons, 56
  health effects, 55-56
  methods of measurement, 67, 69, 73, 85-86
  odor threshold, 292
  solvent extraction system for, 64-67
  see also PAH

321

Arsenic, determination, 157
Arsenite method, for NO_2, 125-128
Asbestos, fibers, 18
  sources and effects, 147
Asphalt, as source of odors, 292
Atomic absorption methods for metals, 152-154
Automated methods of air analysis, 166-195
Automobile exhausts, 61, 98, 148-149

Bakelite, refractive index, 30
Benzaldehyde, 61
Benzene, health effects, 56
  odor, 292, 302
  sources, 62
  ultraviolet wavelength absorption, 68
Benzene soluble organics, analysis, 69-81
  colors and absorption maximum wavelengths, 71
  concentrations by cities, 71-72, 262-263
  fluorescence spectra, 74-76
  health effects, 56, 64
  odor threshold, 298
  solvent extraction systems for, 64-67, 78-79
  sources, 62-63
Benzidine, health effects, 56
Benzo(a)pyrene, 62, 71, 74, 79-81
Beryllium, atomic absorption wavelength settings, 154
  determination, 155, 254-255
  sources and effects, 148
Biological effects of pollutants, 55-56, 200-201
Birefrigence techniques, 27-30
Blood, CO in, 96-97
Bromides, spot test for, 21-22
Bubblers, 122-127, 134, 220-222
  calibration, 124-125, 128, 134-135, 284-286
  see also Absorbers; Absorption methods
Butane, 56, 57
Butanols, odor, 302
Butenes, 57, 61
Butyraldehyde, 61
Butyric acid, odor, 292, 298, 302

Cadmium, atomic absorption wavelength settings, 154
  concentrations in suspended particulates, 262
  sampling for, 252-253
  sources and effects, 149
Calcium, flame test for, 23
Calcium carbonate, analysis, 24-25
  in cement dust, 11
  refractive index, 33
  see also Limestone
Calcium oxide, see Lime
Calcium silicate, in cement dust, 11
Calcium sulfide, refractive index, 32
Calibration, 271-288
  for chromatography, 139-140
  for colorimetric methods, 102, 119-120, 124, 128, 133, 134-135, 136, 184
  of flow-rate meters, 214-216
  for gas chromatography, 88-90
  of high-volume samplers, 240-243, 244
  for Hopcalite method, 105, 106-107
  of hydrocarbon analyzers, 187-188, 190
  of soil-haze samplers, 247-248
Calibration bags, 274, 276. See also Sampling
Calibration gases, see Standard gas mixtures
Cancer, 55-56, 64
Carbon, see Fly ash; Kish; and Spot
Carbon black, 4-5, 50
Carbon dioxide, aqueous solubility, 218
  methods of measurement, 107-108, 258-259
Carbon disulfide, see Sulfur compounds
Carbon monoxide, air quality standards, 268
  biological half-life, 201
  calibration calculations, 279
  diurnal variations, 200
  concentration by city, 262
  health effects, 95-97
  measurement in blood, 96-97
  methods of measurement, 99-107, 170-175, 256-257
  removal from atmosphere, 99
  sources, 97-99
  standard mixtures, 105-107
  statistical evaluation, 263, 265
Carbonaceous dust, see Fly ash
Carbonates, analysis of particulates, 22, 24-25
Carbon tetrachloride, odor threshold, 297

Index 323

Carboxylic acids, as source of odors, 301-302
  separation scheme, 64-65
Carborundum, *see* Silicon carbide
Cellosolve, 63, 82
Cellulose fibers, 18, 21
Cellulose nitrate, refractive index, 30
Cement, birefrigence, 28
Cement dust, 9, 11, 252-253
Centrifugal sampling method, 249
Chemiluminescence methods, 167, 180-181, 185-186
Chlorides, methods for determination, 22, 140-143
Chlorinated hydrocarbons, analysis, 68, 87
  solvent extraction systems for, 64-67
  sources, 62
Chlorine gas, biological half-life, 201
Chromates, spot test for, 22
Chromatographic methods, 66-67, 139-140.
  See also Gas chromatography; Thin-layer chromatography
Chromic acid, sampling for, 252-253
Chromium, concentration of suspended particulates, 262
  spot test for, 22
Chromium oxide, refractive index, 32
Cloud chambers, 192-193
Coal, 2-7
  X-ray diffraction analysis, 46-48
Coke, 7-8
  X-ray diffraction analysis, 46-48
Coke manufacture, as source of contaminants, 62, 109
Collectors, *see* Absorbers; Sampling
Colorimetric methods, continuous, 176-177, 182-184
  for CO, 101-102
  for fluorides, 143-146
  for H_2S, 118-120
  for NO_2, 122-128
  for ozone, 132-135, 182-184
  for SO_2, 110-113, 176-177, 286
  using π acids, 69-72
Color tests, *see* Spot tests, Spectrophotometric methods, Colorimetric methods
Colors, *see* Aromatic colors
Combustion, as source of gaseous pollutants, 57, 62, 98, 109, 120-121, 256-259
  as source of odors, 292
  as source of particulates, 2-9, 256-259
Community air modeling, 249, 254-255, 260-261
Contaminants, inorganic, 95-165
  natural, 1, 62
  odors, 290-291
  organic, 55-94
  particulates, 2-27
Continuous dilution calibration method, 280-281
Continuous methods of air analysis, 166-195
Copper, spot test for, 21-22
Corundum (particulates), 13
Cotton fibers, 19
  birefrigence, 28
Coulometry, 177-178, 182-184
Cresols, solvent extraction systems for, 64
Cresylic acid, sources, 64
Crystal birefrigence, *see* Birefrigence techniques
Crystal structure, *see* X-ray diffraction analysis
Cyclohexane, as solvent for hydrocarbons, 64, 66-67, 76-77

Data interpretation, statistical methods, 261-267
Detergents, 15, 17
Diffraction analysis, *see* X-ray diffraction analysis
Diffusion plate test, 26
Dilution methods of odor measurement, 305-312
Diseases, 55-56
Dispersion staining, 30-34
Diurnal variations in air contamination, 57-60, 98, 138, 198-200
Dolomite, birefrigence, 28
Dry cleaning industry, as source of contaminants, 62, 292
Dry gas meter, 207, 209, 215-216
Dust, *see* Cement dust; Foundry dust; Road dust; and Slag dust
Dustfall, 1-15
  analysis by X-ray diffraction, 43-44
  extraction of organics, 64-65
  particulates in, 9-15

sampling, 2-3, 228-235, 243

Electron microscope techniques, 48-50
Electrostatic precipitation sampling, 248-249
Emphysema, 55-56
Environmental Protection Agency, 52-53
Epoxides, 60, 64
Esters, analysis, 86
   Kovats index, 314-315
   solvent extraction systems for, 64
   sources, 62, 63
Ethane, 57
Ethanol, odor, 302
Ethers, solvent extraction systems for, 64-66
   sources, 62
Ethyl acrylate, odor threshold, 298
Ethylenes, concentration variations in air, 57, 61
   odor, 292, 298
   sources, 62
Extraction, *see* Separation of organics

Federal Ambient Air Quality Standards, 55, 267-268
Fibers, airborne, 15, 18-21
Fick's Law, 282
Filters, 63, 77, 232-235, 237-240, 273
   optimum flow rates, 202
Filtration, for particulate sampling, 232-244
Flame ionization analysis, 58-60, 87, 186-189
Flame photometry, 178-179
Flame test, 23, 67
Flow rate meters, 208, 210-216, 241-243
Flow rates, in sampling, 201-202
Fluorescence analysis, 70, 76-81, 166-167
Fluorescence spectra of common pollutants, 73-76
Fluorides, concentrations by cities, 262-263
   methods for measurement, 143-146, 166-167
   microchemical identification, 26
   sampling for, 254-255
Fly ash, birefringence, 28
   coal, 2-7
   coke, 7-8
      dispersion staining technique, 31
   oil, 6-7
   sampling for, 250-251
Food processing, as source of contaminants, 63, 292
Formaldehyde, 61
   methods of analysis, 83, 258-259
   odor, 292
Fossil fuels, *see* Coal; Combustion
Foundry dust, 11-13
Freeze-out traps, 63, 223-225
Fritted glass absorbers, 122-124, 220-221. *See also* Bubblers
Fuel burning, *see* Combustion
Furfural, sources, 62, 63

Gas analyzers, continuous, 169-186
Gas chromatography, continuous, 173-174
   for odor measurement, 312-315
   for organics, 58, 87-90
   for PAN, 138-139
Gaseous pollutants, analytical methods, 256-261
   sampling, 204, 217-226, 247, 256-261
   see also specific gases, i.e. Carbon monoxide; Nitrogen oxides, etc.
Gas soot, 8
Gas titration method, for ozone, 136, 184-185
Glass, fibers, 18, 30
   fritted, *see* Fritted glass absorbers
   powdered, 13, 16, 47-49
   refractive index, 30
Glycols, Kovats index, 314-315
   solvent extraction systems for, 64
   sources, 62, 63
Grab sampling, 224-226
Gravimetric measurement of suspended particulates, 150
Gravity settling, for dust sampling, 228-232
Griess-Saltzman colorimetry, 122-125
Gypsum, determination by X-ray diffraction, 39-46
   refractive index, 33
   sampling for, 252-253

Halides, spot tests for, 21-22. *See also specific halogen compounds*
Halogens, methods of analysis, 140-146, 258-259
   occurrence, 140-141
   see also specific halogens

Hematite, *see* Iron oxides
Hexanes, concentration variations in air, 57
  photochemical reactivity, 62
High-volume filtration sampling, 235-244, 247
Hopcalite method, 102, 104-107, 174-175
Humidity, effect on odor nuisance, 304-305
Hydrocarbon derivatives, 60-62
Hydrocarbons, air quality standards, 268
  calibration standards, 88-89
  concentration variations in air, 57-60, 262
  Kovats indexing system, 314-315
  methods of measurement, 58-60, 63-90, 186-191, 256-257
  nonmethane, 55, 58-60, 188-191, 268
  odors, 292, 303, 313
  separation of, 57, 63-67, 273, 313-315
  sources, 62-63, 303
  statistical evaluation, 265
  types, 56, 187
Hydrochloric acid, odor, 292
Hydrogen sulfide, aqueous solubility, 218
  biological half-life, 201
  calibration calculations, 287
  methods of measurement, 118-120, 256-257
  odor, 117-118, 292, 298-300
  permeation rate, 282
  sources, 117, 299

Impingers, 112, 221-222, 248
  optimum flow rates, 202
Incineration, as source of contaminants, 9, 10, 109
Infrared absorption analysis, 68
Infrared analyzers for CO, 102-103, 170-173
Instant thin-layer chromatography, 78-79
Iodimetry, 113
Iodine colorimetry, 132-135
Iodine coulometry, 177-178
Iodine pentoxide method for CO, 100-101
Iron, atomic absorption wavelength settings, 154
  concentration of suspended particulates, 262
  determination, 157-158
  microchemical identification, 26
  spot test for, 21-22
Iron oxides, biological half-life, 201
  refractive index, 32, 33
  sampling for, 250-251
  X-ray diffraction spectrum, 45
Isocyanates, sources, 63
Isovaleric acid, odor, 302

Ketones, methods of analysis, 69-70, 82-83, 258-259
  odor, 292
  solvent extraction systems for, 64
  sources, 62, 63
Kish, 12-13
Kovats index system, for odors, 314-315
Kraft process, *see* Paper processing

Lead, atomic absorption wavelength settings, 154
  biological half-life, 201
  concentration of suspended particulates, 262
  flame test for, 23
  sampling for, 252-253
  sources and effects, 148-149
Lead peroxide method for sulfur dioxide, 113-115
Lime, determination by X-ray diffraction, 39-46
  refractive index, 32
  spot test for, 23-24
Limestone, birefringence, 28
  determination by X-ray diffraction, 39-46
  sampling for, 250-251
  see also Calcium carbonate

Magnesium, spot test for, 22
Magnesium oxide, refractive index, 32
Magnetite, *see* Iron oxides
Manganese, determination, 156
Manometers, 210-212, 241-243
Mass spectrography, 58
Measure of undesirable respirable contaminants (MURC), 246-247
Mercaptans, odors, 298, 303
Mercury, biological half-life, 201
  methods of measurement, 154-155, 252-253
  sources and effects, 146-147
  spot test for, 22-23
Mercury vapor method for CO, 175-176
Metals, sources and measure-

326 Index

ment, 146-158, 204
see also specific metals
Meterological variables, 254-255
  effects on odor pollution, 304-305
Methane, concentration variations in air, 57
  determination, 173-174, 186-189, 191
  Kovats indexing system, 314-315
  sources, 62
Methanol, sampling for, 258-259
  sources, 63
Methylene chloride, calibration calculations, 278
  odor, 292
  sources, 62
2-Methyl tetrahydrofuran, odor, 292, 302
Mica, birefrigence, 28
Microchemical techniques for particulates, 25-27
Microscopic identification of particulates, 21-34, 48-52
Molecular structure of odorous chemicals, 292-296
Molybdenum, determination, 156-157
Motor vehicles, *see* Transportation
Mylar, birefrigence, 28

Naphthalenes, aerosol reaction, 63
  sources, 62, 63
National Air Pollution Control Office, 53
National Air Sampling Network, 236, 254
3-NDB test for aromatics, 72-73
Nephelemoter, 190, 192
Nickel, atomic absorption wavelength settings, 154
  concentration of suspended particulates, 262
  sources and effects, 149-150
  spot test for, 21-22
Nitrates, concentration by cities, 262-263
  sampling for, 254-255
Nitric acid, odor, 292
  spot test for, 22
Nitrobenzene, odor threshold, 298
Nitrogen oxides, air quality standards, 268
  aqueous solubilities, 218
  biological half-life, 201
  calibration calculations, 279-280, 286-287
  concentration by city, 262
  diurnal variations, 198
  methods of measurement, 122-129, 167, 181, 256-257
  odor, 303
  permeation rate, 282, 283-284
  sources and chemistry, 120-122, 303
Nondispersive infrared analysis, 170-173
Nylon fibers, 18, 19
  refractive index, 30

Odors, classes, 291-296
  detection and determination, 298-319
  methods of measurement, 258-259, 289-290, 305-316
  sources, 292, 299-304
  theory of, 291-296
Odor thresholds, 117-118, 296-298
Oil burning, particulates from, 6, 7
Oil mists, separation of, 273
Oil refineries, as sources of contaminants, 62, 109, 292, 303
Oils, health effects, 56
  hydrocarbon, sources, 63
Olefins, 56, 57
  methods of measurement, 67-68, 87
Oleic acid, odor, 292
Olfactometer, 289
Olfactory perception, *see* Odors
Optical haze, coefficient of, 245-248
Ore smelting, as source of contaminants, 11-13, 109
Orifice meters, 212, 241-242
Orlon, refractive index, 30
Osmascope, 290
Oxidants, air quality standards, 268
  concentration variations, 199, 262
  measurement, 129-137, 182-186
  see also Ozone
Oxides, *see specific oxides, e.g.,* Zinc oxide; Iron oxide; Silicon dioxide
Oxygenates, 60-62
  analysis, 68, 81-87
  odor threshold, 298
  solvent extraction systems for, 64-67
Ozone, 129-137
  mechanisms of formation, 130-131
  methods of measurement, 131-137, 167, 182-186, 256-257

PAH, determination, 73-81
Paint, as source of contaminants, 63, 149, 292, 303-304

Index 327

PAN, 131
  methods of measurement, 138-140, 260-261
  occurrence, 61-63, 137-138
Paper processing, as source of contaminants, 302-303
Paracresol, odor threshold, 298
Particle Atlas, 52
Particulates, air quality standards, 52-53, 268
  analytical methods, 20-52, 116, 156, 157, 250-255
  atomic absorption spectrophotometry, 152-154
  concentrations by cities, 262-263
  continuous monitoring, 190, 192-193
  determination of acidity or basicity, 150
  gravimetric measurement, 150
  metals, 22, 147-158
  sampling, 204, 226-255
  separation of, 273
  size and shape, 48-50, 53, 227, 233
  sources and identification, 1-54
  suspended, 53, 150, 190-193, 232-249, 262-263, 268
  X-ray diffraction analysis, 34-48, 50-51
PBN, 61, 63
Pentanes, concentration variations in air, 57
Permeation tubes, 272, 281-286
Peroxide, 129-130
Peroxyacetyl nitrate, *see* PAN
Peroxyacyl nitrate, *see* PAN
Peroxybenzylnitrate, *see* PBN
Petroleum, *see* Oil refining
Petroleum process catalyst, 15, 17
Pharmaceutical industries, as sources of hydrocarbons, 62
Phenolphthalin method, for ozone, 137-138
Phenols, determination, 69-70
  odor, 292, 298, 303
  sampling for, 260-261
  solvent extraction system for, 64
  sources, 62
Phosphates, 15, 17
  microchemical identification, 26
  spot test for, 22
Phosphoric acid, odor, 292
Photochemical smog, air quality standards, 55, 268
  formation, 55-56, 58-60
Photoreactive hydrocarbons, 57-62
Phthalic anhydride, sources, 63
Piperonal test, 70-72
Pitot tubes, 213-215
Plastics industries, as sources of hydrocarbons, 63
PNAH, *see* PAH
Pollutants, *see* Contaminants
Polycyclic aldehydes, analysis, 85-86
Polycyclic aromatic hydrocarbons, 57
  health effects, 55-56
  methods of measurement, 67, 69, 73-81
  solvent extraction systems for, 64-67
  sources, 76
Polycyclic compounds, sampling for, 254-255
Polymer fibers, 18
Polynuclear aromatic hydrocarbons, *see* PAH
Polystyrene, refractive index, 30
Potassium, flame test for, 23
Potassium iodide method for ozone, 132-135, 182-184
Power generation, as source of contaminants, 109
Propane, health effects, 56
  sources, 62
Propanols, odor, 302
  sources, 63
Propene, 61
Pulp processing, *see* Paper processing
Pumps, for sampling, 205-206
Pyrenes, fluorescence spectra, 74-76
  methods of analysis, 67-73
  solvent extraction systems for, 64-67
  sources, 62
Pyridine, odor threshold, 298

Quartz, *see* Silicon dioxide

Rayon fibers, 18, 20
Reactive surface sampling method for sulfur dioxide, 113-115
Refractive index techniques, 27-34, 67
Rendering odors, 292, 300-302, 309-311
Refuse incinerator particulates, 9, 10
Ring oven technique for metals, 151-152
Road dust, 9
Road salt, 30, 252-253

328 Index

Rotameters, 212-214, 237, 241-243
Rubber cracking method, for ozone, 136-137
Rust, refractive index, 33

Salt, *see* Road salt; Sodium chloride
Sampling, 197-270
   for collection, of inorganics, 122-124, 127, 134, 140-142, 144
   of organics, 63-67, 77
   of particulates, 2-3, 53, 152-153, 226-255
   continuous, 166-169
   equipment and devices, 123, 127, 134, 205-216, 220-226, 235-238. *See also* Sampling instruments
   for gases, 217-226, 256-261
   general requirements and procedures, 196-205
   for odor measurement, 305-307, 313
   statistical methods, 249, 254-255, 260-267
   units of measurement, 204
   see also specific analytic methods
Sampling bags, 226
Sampling instruments, calibration, 271-288
   continuous, 167-169, 271, 280-285
Scentometer, 316
Sedimentation, *see* Settling rates of particles
Selective solvent systems, *see* Solvent systems
Selenium, determination, 155-156
Separated organic fractions, analysis, 67-68
Separation, for microchemical analysis, 26-27
   of organics, 63-67, 73, 77-79
Settling rates of particles, 227
Sewage incinerator particulates, 9
Silica, *see* Silicon dioxide
Silicon carbide, 13-14
   birefrigence, 28
   refractive index, 33
Silicon dioxide, birefrigence, 28
   determination by X-ray diffraction, 40-46
   particulates, 13, 15
   refractive index, 33
   sampling for, 250-251
   X-ray diffraction analysis, 40-42, 44-45
Slag dust, X-ray analysis, 48, 49

Smog, *see* Photochemical smog
Sodium, flame test for, 23
Sodium chloride, refractive index, 30, 32
Soil-haze indices, 246
Soil-haze particle sampler, 243, 245-248
Solubilities, *see* Aqueous solubilities
Solution absorption, *see* Absorbers; Absorption methods
Solvent degreasing, as source of hydrocarbons, 62
Solvent systems for extraction of organics, 64-67, 77-79
Soot, 3-7, 8
   health effects, 56
   sampling for, 250-251
Spectrophotofluorometric method, 73-76, 79-81
Spectrophotometric methods, for CO, 102-104
   for fluorides, 145-146
   for hydrocarbons, 67-68, 69-71
   for metals, 152-154
   using π acids, 69-72
Spirometer, 206-208
Spot tests, 21-24, 70-71
Standard gas mixtures, 88-89, 105-107, 139-140, 271-288
Standards, for air quality, 52-53, 55, 147, 267-268, 300
   for gas chromatography, 88-90, 139-140
   for X-ray diffraction analysis, 43-45
Starch, as source of odors, 292
   refractive index, 30
Statistical methods, use in sampling, 249, 254-255, 260-267
Stoke's law, 227, 235
Storage containers for gases, 273-274
Strontium, flame test for, 23
Substituted hydrocarbons, analysis, 81-87
Sucrose, birefrigence, 28
Sulfates, aerosol analysis, 116
   concentrations by cities, 262-263
   sampling for, 254-255
   spot test for, 22
Sulfation rate, 115
Sulfur compounds, odor, 292, 298-300, 302, 303
   sources, 62, 63
   spot test for, 22
Sulfur dioxide, air quality standards, 268

biological half-life, 201
calibration calculations, 278-279, 283, 286, 287
concentration by city, 262
concentration hourly, 263-264
methods of measurement, 110-117, 175-179, 202-203, 256-257
permeation rate, 282
sources, 108-110
Sulfur oxides, *see* Sulfur dioxide
Sulfur slag, *see* Calcium sulfide
Sulfuric acid, as source of contaminants, 109
odor, 292
Sulfuric acid mist, sources and determination, 115-117, 250-251
Suspended particulates, *see* Particulates, suspended
Syringe method for odor measurement, 305-307

Tallow rendering, *see* Rendering odors
Tape sampler, 243, 245-248
Tars, 56, 64, 68
Terpenes, odor, 292
Textile industries, as source of hydrocarbons, 62
Textiles, *see* Fibers
Thermal precipitation sampling, 248
Thin-layer chromatography, 73-79
Thiols, sources, 62, 63
Thiophene, odor, 292
sources, 62
Titania, refractive index, 33
Titanium, spot test for, 21-23
Titration methods, 141-143, 146. *See also* Gas titration method
Titrimetry, for determination of sulfur dioxide, 113
Toluene, odor, 292, 298
ultraviolet wavelength absorption, 68
Transportation, as source of pollutants, 62, 121. *See also* Automobile exhausts
Trimethylamine, odor threshold, 298

Ultraviolet absorption spectrophotometry, 67-68
Ultraviolet light, absorption by atmospheric components, 67-68, 130-131
Urban atmosphere variations, 58-60, 71-72, 138, 262-264

Vanadium, spot test for, 21-22
Vinyl chloride polymer, refractive index, 30

Water vapor, separation of, 171-172, 272-273
Wavelength absorption maxima of organics, 68, 71
Weather, *see* Meterological variables
West-Gaeke colorimetric method, 110-113, 176-177, 179
Wet test meters, 208-209
Windblown soil, 9-10
Wood fibers, 18
Wool fibers, 20

X-ray diffraction analysis, 34-48, 50-51
X-ray target materials, radiation wavelengths, 45-46
Xylenes, photochemical reactivity, 62
ultraviolet wavelength absorption, 68

Zinc, atomic absorption wavelength settings, 154
spot test for, 22-23
suspended particulates, 262
Zinc oxide, particulates, 13-14, 250-251
refractive index, 33